The Generation of Business Fluctuations

Dynamische Wirtschaftstheorie
Dynamic Economic Theory

Edited by/Herausgegeben von Carl Chiarella, Peter Flaschel, Reiner Franke, Michael Krüger, Ingrid Kubin, Thomas Lux, Willi Semmler, Peter Skott

Vol./Bd. 26

PETER LANG

Frankfurt am Main · Berlin · Bern · Bruxelles · New York · Oxford · Wien

Corrado Di Guilmi

The Generation of Business Fluctuations

Financial Fragility and Mean-field Interactions

PETER LANG

Internationaler Verlag der Wissenschaften

Bibliographic Information published by the Deutsche Nationalbibliothek
The Deutsche Nationalbibliothek lists this publication in the Deutsche Nationalbibliografie; detailed bibliographic data is available in the internet at <http://www.d-nb.de>.

ISSN 0930-2174
ISBN 978-3-631-58119-3

© Peter Lang GmbH
Internationaler Verlag der Wissenschaften
Frankfurt am Main 2008
All rights reserved.

Printed in Germany 1 2 3 4 5   7

www.peterlang.de

PREFACE TO THE SERIES

The editors of 'Dynamische Wirtschaftstheorie' wish to encourage and support contributions to dynamic economic analysis. The series is open to contributions in both macro- and micro-economics, including work on business cycles, economic growth, the stability of market systems, and the dynamic interaction between heterogeneous agents. The application of new methods in the theoretical or applied analysis of economic dynamics is particularly welcome.

Manuscripts should be no less than 100 and no more than 400 pages in length. The preferred language is English, but submissions in German, French, Spanish and Italian will be considered too. Conditions for the publication of dissertations can be obtained from the editors. Please send all submissions (2 copies) to the coordinating editor:

Professor Peter Flaschel
Bielefeld University
Faculty of Economics
Postfach 100 131
D-33501 Bielefeld
Germany
email: pflaschel@wiwi.uni-bielefeld.de
Phone: 0049 - (0)521-106 5114 / 6926
Fax: 0049 - (0)521-106 6416

Acknowledgments

First of all I want to thank Prof. Mauro Gallegati. Without his trust I would not even have begun to do research. Perhaps science would not have suffered an irreparable loss, but I will never forget all the experiences and opportunities that the our cooperation has offered to me.

Special thanks go to Simone Landini of the IRES Piemonte for his patience, support and friendship. Without him, reading Aoki's books in English or Japanese would not made a big difference to me. He can be considered as a co-author for some analytical parts of this work, even though responsibility for all eventual errors remains mine.

My gratitude goes also to, in a strictly random order: Tiziana Di Matteo, Tomaso Aste and all the friends of the Department of Applied Mathematics of Australian National University; Fabio Clementi, Marco Gallegati and Giulio Palomba of the Università Politecnica delle Marche; Vasilis Sarafidis of the University of Sydney; Sean Holly and Steffan Ball of the Cambridge University; Yoshi Fujiwara of the ATR Japan; Steve Keen of the Western Sydney University; Antonio Palestrini of the University of Teramo; Domenico Delli Gatti of the Catholic University of Milan; Carl Chiarella of the University of Technology, Sydney; Edoardo Gaffeo of the University of Trento.

I cannot forget the support given by my parents, since without them I would often have slept under a bridge and it would have been at least uncomfortable to write the thesis under a bridge.

A very particular thank you goes to Giulia, Cristiano and Daniele. This work is dedicated to them.

Contents

List of Figures

List of Tables

Chapter 1

Introduction

1.1 Aim and results

In the last decades, the complex system theory gained increasing attention from economists[1]. Relevant advances have been recently reached with the generation and development of models that import concepts and tools from hard sciences into economics. They represent one of the attempts to overcome the unsound foundations of mainstream tradition (Keen, 2001) and, in particular, to substitute the representative agent framework, that leaves substantially unsolved the problem of aggregation and, then, of the link between micro and macro (Kirman, 1992). A system can be defined as complex when it is populated by a large number of components that interact, giving rise to nonlinear feedback effects and displaying, at aggregate level, emerging properties and self-organization phenomena that cannot be directly inferred by components' behavior[2]. Therefore, in a complexity perspective, an economic model should be composed by heterogeneous agents that interact among them and with a meso and a macro level. Considering the nonlinearity and the uncertainty that is inherent in such a structure, it should be formulated in stochastic terms.

The main contributions in this analysis moved, until now, towards two basic directions: on one hand, the development of agent based models, treated and numerically solved by means of computer simulations[3]; on the other, the implicit formulations of complex or stochastic mechanic frameworks in terms of (micro and macro) economic variables[4]. No attempts have been made to analytically find an explicit and close solution for this kind of models, as

---

[1] Arthur (1999); Anderson et al. (1988); Rosser (2004) among others.
[2] Bak (1997); Rocha (1999).
[3] Axelrod (1997); Axtell et al. (1996).
[4] Aoki (1996, 2002); Aoki and Yoshikawa (2006).

commonly performed in traditional economic analysis. Indeed, if a growing number of models (partially or wholly) accept the complexity paradigm, the individuation of aggregate results is reached either through computer simulations or imposing restrictions on agents' behaviors. On the other hand, the problem of how to sum up heterogeneous and evolving agents cannot be handled with the usual tools of the economist and the aggregation methods that have been proposed are still waiting for an application that, using explicit functions and formulations, succeeds to demonstrate their usefulness.

This work is meant to be a step in filling this gap. The paper of Delli Gatti et al. (2005) imported the well known Newkeynesian bankruptcy approach into a complexity perspective, introducing heterogeneous agents in the model of Greenwald and Stiglitz (1993). As demonstrated in this work, a structure of heterogeneous agents appears to be particularly suitable for an implementation in a financial fragility approach, where, by construction, a representative firm cannot represent bankrupted firms and, given the possibility of different financial structures, the distribution of agents matters (Forni and Lippi, 1997). Minsky's Financial Instability Hypothesis, since its first formulation (Minsky, 1963), distinguishes firms on the basis of their financial soundness.

At the same time, Masanao Aoki managed to remove the representative agent hypothesis, introducing in economics the ground-breaking concept of *mean-field interaction*, that makes the analytical aggregation of heterogeneous agents feasible, replacing the unrealistic mechanic determinism of mainstream framework with a set of stochastic tools borrowed from statistical mechanics. In particular, he adopted the *master equation* to describe the dynamics of probability flows, and, thus, to determine the aggregate effects of underlying fluctuations at micro level.

In this book, I start from these two contributions to develop a macroeconomic model of financial fragility, in which the interaction among heterogeneous agents gives rise to statistical regularities at macro level. According to mean-field modeling, firms are grouped in a pre-defined set of states, according to their financial structure. It's only the evolution of the number or proportion of agents that occupy these states that determines the dynamics of aggregate variables. To mimic the inherent uncertainty of real world, the occupation numbers are governed by a stochastic law that defines also the functional of the probability distributions of aggregate variables and, if existing, their equilibrium distributions. Interaction is made effective through feedback loops that influence the stochastic mechanism.

The main result is the development and the use of a framework for the aggregation of heterogeneous agents that demonstrates to be capable of originating fluctuations of total production around a long path trend. In particular, finding a steady-state solution for the master equation, it is possible to identify two stochastic equations in order to describe the dynamics of the trend and of the fluctuations. The aggregation procedure is explicit and mathematically demonstrated and it does not need any heroic hypothesis or particular functional formalization of agents' behavior. The macro level outcomes are obtained without the formulation of ad hoc-hypothesis about the dynamics of the system or about its starting (or final) conditions. Consistently with the inspiring approach, aggregate output is an inverse function of the degree of financial fragility of the system. It is also shown that a policy that does not consider the economic system as complex has little or no chance of modifying its statical or dynamic conditions.

In a certain sense, there is another gap that this work could partially help to fill. At its origin, economics borrowed its analytical structure from physics, adopting the Newtonian reductionism that was the dominant paradigm at that time. Then, after the quantum revolution, new approaches such as statistical mechanics and, later, complexity theory developed and became available for application in social sciences, for which they seem to be more suitable. But economics, in the meantime, cut its umbilical cord and did not (want to) realize it, determining an evolution of the discipline that led it away from its object, in a sometimes dramatic way. Today the effects of the belief in a spontaneous equilibrium are manifest. In a world where three quarters of humanity starves, a little more realism seems to be more than a scientific eccentricity.

Finally, a word of caution. The adopted mathematical tools will likely appear computationally quite burdensome. In my opinion, this impression may be, at least partially, caused by the novelty of the methods for most of the profession, in comparison to the mathematical foundation of neoclassical economics that have been largely assimilated. In the Preface to *The General Theory*, Keynes wrote:

> "The difficulty lies, not in new ideas, but in escaping from the old ones, which ramify, for those brought up as most of us have been, into every corner of our minds".

1.2 Outline of the work

In the first introductory part, I depict, in chapter 2, the methodological (section 2.1) and substantial (section 2.2) contexts in order to properly situate the original contribution of the work. Chapter 3 provides a brief insight into the key instruments exploited in the following part. The second part is devoted to the presentation of the model, starting, in the two sections of chapter 5, from the detail of the foundation hypothesis about the stochastic structure and about the adaptation of the assumptions of original models. Then, in chapter 6 stochastic inference is performed, developing the analytical solutions. Section 6.1 develops the estimation method and the solution for statical probability functions. Section 6.2 is the central part of the work, where the method of solution is introduced and developed, ending with the determination of equilibrium distribution and dynamic solutions of the model. Chapter 7 sums up the results, linking the analytical solutions developed in previous chapter to firms behavior, completely defining the analytical structure. This second part ends with two applications of monetary policy (chapter 8). The third and last part of the book presents different empirical evidence. First (chapter 9), the model is applied on simulated and real data, in order to test its informative power. Then chapter 10 presents a wide set of evidence about business cycles and its relation with debt, along the lines of Minsky's Financial Instability Hypothesis. Finally, the relevance of firms' financial structure is further evidenced in chapter 11, where an experimental implementation of engineering tools to estimate the expected profitability of firms is proposed.

Part I

The context

Chapter 2

From micro to macro

2.1 The aggregation issue

The first distinction among microeconomics and macroeconomics can be dated back to Lindahl's essay of 1919, that was later included in a collection (Lindahl, 1939), in which he theorized an interaction among individuals that can generate complex social formation as emergent phenomena. A picture that appears quite distant from the dominant present distinction, in which the representative agent hypothesis lets microeconomics actually subsume macroeconomics. Macroeconomics became, basically, micro spoken in a louder voice.

The representative agent hypothesis has been introduced, substantially, as a short cut to avoid the problem of *exact aggregation*. Gorman (1953) demonstrated that demand functions of single consumers can be aggregate, independently from agents' distribution, only if they are parallel hyperplanes in individual variables: the well known *polar forms*. They are the result of homotetic individual preferences that determine the same structure of marginal preferences for all consumers. In absence of this condition aggregate demand curve could show kinks or even be positively sloped (Keen, 2001). It actually means that all the consumers show the same response to a variation in price. In order to get this result two implicit hypothesis have to be stated: first, the distribution of income does not matter; second, the distribution of income does not change after a variation in relative prices.

To overcome (or to better hide) these, at best, restrictive hypothesis, it has been theorized that economy is populated by identical individuals, so that, studying the single agent's behavior, we can reproduce the aggregate, simply correcting by a scale factor. This framework is at odds with empirical evidence (Stoker, 1993) and presents a series of drawbacks. It is evident that, if all individuals are alike, there is no reason to trade, as demonstrated by the so-called "no-trade" theorems (see Rubenstein, 1975, among many others).

Kirman, in his influential paper of 1992, analyzes other four weaknesses for which representative agent hypothesis "*is not only primitive, but fundamentally erroneous*". First, there is no justification in assuming that the maximization of a single's utility implies the maximization of social utility. Second, analogously to Lucas' critique, what happens to this representative individual when an exogenous modification takes place? The researcher should construct another and different representative agent, with no guarantee that his choices remain unchanged. As a result, there is no way to correctly estimate the reaction, for example, to a policy intervention. Third, since all individuals have the same marginal preferences, there is no room for the representative individual to disagree with others. In the not unlikely eventuality that one has different preferences from another[1], how then is the representative agent chosen? Finally, in order to perform empirical tests of a model, one has to introduce simplifications, rejecting some behavioral hypothesis. To quote Kirman:

> "if one rejects a particular behavioral hypothesis, it is not clear whether one is really rejecting the hypothesis in question, or rejecting the hypothesis that there is only one individual".

From an econometric perspective, Palestrini (2000) shows how the attempt of Theil (1954) to estimate Gorman's forms fatally incurs in a specification error. Modeling linear functional forms, Theil omits all microvariables of the aggregate function and substitutes them with macrovariables, for what he defines as the *analogy principle*. But macrovariables are not independent from microvariables, since the former are simply the sum of the latter, and, as a consequence, he uses correlated variables.

Extensions and an application of exact aggregation in less restrictive conditions are proposed by Jorgenson et al. (1982) and Lewbel (1992). In both works aggregation is obtained by means of a statistic of a transformation of microvariables. In particular, Lewbel (1992) proposes a log-linear model for the aggregation of the mean values of firms' distribution. The conditions for the estimation of the model appears, also in this case, very restrictive. In particular, to aggregate a series of j micro-relations, referring to the characteristic i, of the type:

$$ln(y_{ijt}) = \sum_{j=1}^{J} b_j ln(x_{ijt}) + r_{ijt}$$

---

[1]It is enough that this is verified for just one good.

in order to obtain the aggregate relation:

$$ln(Y_t) = \sum_{j=1}^{J} b_j ln(X_t) + R_t$$

it needs that the elasticities b are the same between aggregate and singles for each possible value of b. It is shown that this condition is reachable if the variable is *mean scaled*, i. e. the concentration of firms is independent from its average value, without any hypothesis about their distribution. Formally, the distribution $G(z, |X_t\theta_t)$ of the variable $z_{ijt} = x_{ijt}/X_{ijt}$, for the aggregate parameter θ_t, has to be independent from X so that:

$$G(z|X_t, \theta_t) = G^*(z|\theta_t)$$

It appears clear that e. g. a policy intervention on income is likely to change the mean of a distribution and, therefore, to distort aggregation results.

The distribution at micro level has to be taken into account in order to obtain consistent aggregation procedures, considering, at the same time, the modifications that functional forms exhibit among micro and macro level. A first example of *stochastic aggregation* is the one of Stoker (1984), that allows to introduce heterogeneity among micro relations and between micro and macro functional forms, conditioned on the existence of a joint probability distribution of microvariables. He makes use of the first moments of this joint probability distribution, obtaining a formulation for the aggregate expected variable that is a function of a subset of its parameters and of the expected values of microvariables. Formally, having the ith microrelation:

$$y_{it} = f_i(x_{it}, u_{it}, \theta_i) \tag{2.1}$$

where y is the strategic choice and x are the observables, with joint probability function:

$$y_{it} = f_i(x_{it}, u_{it}, \theta_i; \phi_t) \tag{2.2}$$

he estimates a macro relation equal to:

$$\mu_{yt} = \Psi_y \left[\phi_{1t}, \Psi_x^{-1}(\phi_{1t}, \mu_{xt})\right] = F(\mu_{xt}, \phi_{1t})$$

where ϕ_{1t} is a subvector of parameters of the joint probabilities and $\phi_{2t} = \Psi_{(.)}^{-1}(\phi_{1t}, \mu_{(.)t})$.

An explicitly alternative approach is following by Gallegati et al. (2006), who introduce the *variant representative agent*. They use Stoker's method and combine it with Aoki's approach. They expand in Taylor series equation (2.1) obtaining an approximation of the first order and an approximation of the second order. Inverting these functions, they are able to estimate the first and the second moment of joint distribution. The micro-macro relationships are then dynamically modeled as a system of coupled equations by means of which it is possible also to estimate relations among macrovariables. This approach represents a clear cut with representative agent framework and appears quite promising for further applications.

Since the core of this book is an application of Aoki's methods, for the completeness of the review, I just anticipate here some basic features. Relatively to the frameworks presented so far, the dynamic stochastic aggregation procedures proposed by his recent works appear to be, on one hand, the most generically applicable, since they do not need any particular specification for the underlying model, and, moreover, the most informative, since, with few inputs, they can return estimations of aggregate distribution probabilities at each point in time, permitting, in this way, a fully dynamic and complete modeling of the economy as a complex system. Two main components of his approach are identifiable:

1. continuous time Markov chain to model stochastic dynamic interactions among agents;

2. combinatorial stochastic processes, i. e. the combination of stochastic process and non-classical combinatorial analysis.

These components are defined in chapter 3. For the moment it will be enough to specify that, as regards point 1, by means of a *mean-field interaction* structure, it is possible to model the stochastic behavior of interacting heterogeneous agents by grouping them in clusters. The dynamic evolution of the system is described by a *master equation*. The second component describes the stochastic dynamic processes of formation of clusters of agents, the probability distribution of the size of these clusters and the probability of access to each of them. Economy is a complex system populated by a very large number of agents (e. g. in Italy there are about 10^6 firms) and therefore we cannot know *which* agent is in *which* condition at a given time and if an agent changes his condition, but we just need to know which is the present probability of a given state of the world. In this theoretical context, to quote Aoki and Yoshikawa (2006),

> "precise behavior of each agent is *irrelevant*. Rather we need to recognize that microeconomic behavior is fundamentally stochastic and we need to resort to proper statistical methods to study the macroeconomy consisting of a large number of such agents".

This assertion seems to echo Keynes when he writes:

> "It is sometimes said that it would be illogical for labour to resist
> a reduction of money-wages but not to resist a reduction of real
> wages. For reasons given below, this might not be so illogical
> as it appears at first; and, as we shall see later, fortunately so.
> *But, whether logical or illogical, experience shows that this is how
> labour in fact behaves*" (Keynes, 1936).

While the variant representative agent framework is yet too young to
find application in literature, Stoker's method, mainly for the necessity of a
lot of information on micro-level distributions (Pesaran, 2000), is not widely
diffused among the profession. As anticipated in chapter 1, this book, using
Aoki's theories, represents a first attempt to provide a complete application
of one of the above mentioned aggregation methods completely alternative
to the representative agent framework.

2.2   Financial fragility in the New Keynesian approach

The links between the financial variables and macroeconomic fluctuations
have been brought to the attention of economists by the clamorous failure
of Say's law that took place for the Great Depression. As known, Say's
law states that "supply creates its own demand" and therefore, like most
economists at that time, it does not contemplate the possibility of crisis for
insufficient demand. The explanation advanced by Keynes (1936) involved
the uncertainty, that is inherent to economic systems, and the role of money,
that was revealed to be something more than a simple "veil".

The *debt deflation school* (Fischer, 1932), developed a theory about the
mechanism that from defaults on debt payments leads to a decline in aggregate demand and, as a result, of a reduction in prices. The consequent
increase in the real value of debt payment commitments determines an interacting downward spiral. Later, Minsky (1963), formulating the *Financial
Instability Hypothesis*, presented a systematic approach to the analysis of how
a financial crisis can affect the real sector of the economy. According to his
theory, instability is unavoidable in a capitalist economy due to its dependence on credit. Firms can be distinguished, on the basis of their short-term
financial structure, in three types: *hedge*, i. e. firms that can meet short-
term liabilities with cash flow; *speculative*, that may have to re-finance current liabilities; *Ponzi*[2], that cannot even meet the payment of interests, and

---

[2]From Charles Ponzi, the inventor of the so-called financial pyramids that caused to
him some short periods of wealth and longer periods of detention.

therefore, continue to rise their outstanding debts. During expansions, banks grant credits with growing facility, since the rise in the level of economic activity spreads a sort of contagious optimism. The level of debt considered as acceptable grows, as expansion makes the banks confident about repayment. This confidence leads to a boom in economic activity, that contains in itself the germ of the crisis since the proportion of Ponzi firms have augmented in the meantime. When the Ponzi firms begin to fail, liquidity declines and banks start to reduce lendings, causing a stagnation that becomes depression as financial distress extends to other firms.

As Minsky himself pointed out, a crisis such as the Great Depression in modern economies may be avoided by the presence of a *big Bank*, i. e. a central bank that acts as a lender of last resort, and of a *big Government* that sustains demand when it falls. But recently, Financial Instability Theory came back into fashion for two main reasons. The first is the relevance assumed by emerging economies in countries that have neither a big Bank, nor a big Government; the Asian crisis can, in this sense, be interpreted as a confirmation, in the negative, of Minsky's intuitions (Kregel, 1998). The second is represented by the fact that Minsky wrote in a period when there were great barriers to the circulation of capital. These barriers have now almost completely disappeared, leaving all economies potentially exposed to international crisis. While I'm writing, the sub-prime mortgages crisis, spread in USA, has not yet manifested all its effects but has started a sense of panic that leaded, for the first time, the Governor of the Bank of England to solemnly confirm the role of lender of last resort of the institution. Empirical evidence on these aspects is reported in section 10.3.

The correlation between financial variables and aggregate fluctuations gained new consideration from the 80s for the interest of New Keynesian economics in the analysis of markets with asymmetric information (Stiglitz, 1982). Specifically, in New Keynesian macroeconomic models, the reason of rigidities in nominal values can be found in the informational imperfections in real (Mankiw, 1985) and financial (Greenwald and Stiglitz, 1993) markets. For financial markets in particular, the presence of informational asymmetries among firms and investors gives rise to two phenomenon, that have been put into evidence by this stream of literature: the emerging of a hierarchy of preferences among the source of financing (Myers and Majluf, 1984), contrarily to what was postulated by the Modigliani-Miller theorem (1958), and the rationing of credit (Stiglitz and Weiss, 1981) and equity (Greenwald et al., 1984) for firms. The macroeconomic consequences of this kind of informational asymmetries have been analyzed in different analytical perspectives: bankruptcy cost approach (Greenwald and Stiglitz, 1993), agency cost approach (Bernanke and Gertler, 1989), and the approach followed by Kiyotaki and Moore (1997), based on assets and collateral securities.

As anticipated in the introduction, this work follows a bankruptcy approach, as modeled in Greenwald and Stiglitz (1993). In this analytical structure, macroeconomic fluctuations are due to the rationing in equity market that obliges firms to finance investment with new debt, exposing them to the risk of default. To solve the model in a representative agent framework, the authors are forced to assume that all firms use all their net worth and then recur to debt. In this way, they assume, in fact, the existence of one only marginal source of financing, in order to have an equal marginal cost of debt for each firm and then treat them all as identical. As sketched above, according to Myers and Majluf (1984), financing sources have different costs for firms and these differences originate a hierarchy in their preferences. Therefore, the introduction of the option for firms to finance all investments with equities, and consequently, of the possibility of two different marginal costs, makes the assumption of identical firms inconsistent. As exposed in chapter 4, Gallegati (2002), introducing a second source of financing in this original structure, considers a system with firms that, following a binomial distribution, are divided in two classes.

More in general, in the case of a systemic financial fragility depending on balance sheet variables, it is the distribution of agents that, far from being neutral, determines business cycles. If, as theorized by Minsky, firms are different for financial structure, there will be different risks of default and, then, differences in marginal costs of financing. The absence of proper statistical tools for aggregation or the need to speak in a codified language to stress the macroeconomic result leads to a simplification of the scenario to make model's solution feasible.

As exposed above, the aim of this work is to develop a solution of an analogous model without the restrictive hypothesis in matter of financing of Greenwald and Stiglitz (1993) and without the pre-defined assumption of statistical distribution of agents introduced by Gallegati (2002). Specifically, the framework presented in part II appears to be more consistent with a financial fragility approach where, on the one hand, the distribution of agents determines macroeconomic properties of the system and, on the other, this distribution is endogenously determined by the system itself.

Chapter 3

Some basic definitions

In what follows the basic mathematical instruments exploited in chapter 6 are briefly presented. In particular, the first three sections treat the basic features of the statistical structure on which the model is built (section 3.1) and the dynamic instruments for the functional definition of probabilities fluxes (section 3.2) and for the identification of stationary state solutions (section 3.3). Section 3.4 presents the meaning and the measures of statistical entropy and the method of solution of entropy maximization problems, that will be used for the quantification of statical probabilities.

3.1 Markovian and Gibbs random fields. Mean-field interaction

Given a probability space $(\mathbf{U}, \mathcal{F}, \mathbf{P})$, a *stochastic process* is a collection, in the state space $U \in \mathbf{U}$, of random variables indexed by a set T. Being $t \in T$ a time index, the collection can be defined as a time series, with path U. All the possible paths define a tree structure. Thus, a stochastic process F is a collection:

$$\{F_t; t \in T\}$$

where each F_t is a function with domain the sets of possible paths $\mathbf{U}$ and value the outcome at time t.

A stochastic process whose domain is a region of space (instead of a subset of time) is a *random field*. Thus, a random field can be defined as an ensemble of random numbers mapped into a space (lattice). There exists a (spatial) correlation among them, and, in its most basic form, this correlation exists only (or results higher) for adjacent constituents. This structure appears to be suitable to provide a formal representation of stochastic interactions among different elements.

Follmer (1974) first applied random field theory for interaction in economics, building a model in which agents' preferences can be influenced by interaction among them[1]. The interaction structure is modeled in Markov random fields, since they display some properties that make it particularly treatable. A random field, defined in a space Ω, is classified as markovian if its distribution or its density function displays the following properties:

1. positivity condition: $P(X = x) > 0 \; \forall x \in \Omega$. So that the system cannot remain locked in a state;

2. Markov property:

$$P(X_i = x_i | X_j = x_j, \forall j \neq i) = P(X_i = x_i | X_j = x_j, j \in N(i))$$

   where $N(i)$ is the ensemble of neighborhood of i. It states that each agent is influenced (interacts with) only by his neighborhoods;

3. homogeneity condition:

$$P(X_i = k | X_j = x_j, \forall j \neq i) = P(X_s = k | X_j = x_j, j \in s)$$

   that is to say that every cell has the same probability functional form.

Let me now introduce Gibbs random field. For our purpose it will be enough to say that a Gibbs random field is a measure of probability defined so that the expected stationary probability can be expressed by:

$$P(X = x) \propto Z^{-1} e^{-\beta U(x)} \tag{3.1}$$

where $U(x)$ is the *Gibbs potential* (Woess, 1996) and can be defined as a functional of the local dynamic characteristics of the control variable. In a system with N fields, the equation of the potential is:

$$U(x) = -N \int_0^x g(x) dx - \frac{1}{\beta} H(\underline{X})$$

where $H(\underline{X})$ is the Shannon entropy, that will be defined in section 3.4, and $g(x)$ is a function that evaluates the relative contribution of each *clique*, i.e. a system of neighborhood on the lattice.

Mean-field interaction can be defined as the average interaction model that substitutes all the relationships among agents that, otherwise, could

---

[1]Other successive applications can be found in Arthur et al. (1987); David and Foray (1993).

not be treated analytically (Opper and Saad, 2001). To get a quantitative clue, in this kind of model, interaction influences the mean of the state variable, indeed it is also knows as mean-field approximation. In chapter 6, the effect of mean-field interaction are quantified (in probability) by the potential, applying the tools presented in this chapter.

3.2   Master equation

A complete exposition of master equation's derivation goes far beyond the aims of this book. I just sketch here the basic notions that will permit, in section 6.2, the possible clearest exposition of its use in the theoretical model[2]. As a first approximation, master equation can be defined as a phenomenological first-order differential equation, that describes the dynamics of the probability of a system to occupy each one of a determinated set of states. States can be defined as sets of information that are sufficient to determine the evolution of probability distribution of the system's configurations, given all the other information on external influences that may affect model behavior (Aoki and Yoshikawa, 2006; Bellman, 1961). In the Markov model states may be subsets in Euclidean spaces or graphs.

Let me preliminarily define a stochastic Markov process as a finite stochastic process that displays Markov property, so that the nth value depends only on the $(n-1)^{th}$:

$$P(x_t|x_0, ..., x_{t-1}) = P(x_t|x_{t-1})$$

Every continuous time Markov process is associated with an embedded Markov chain, that is defined as a jump process in discrete-time, which displays Markov property[3]. It is fully described by the transition probability matrix (Ross, 2003) with entries p_{ij}, that represent the conditional probabilities of transitioning from state i into state j. In the homogeneous case, transition probabilities $p_{ij}(t)$, do not depend on time and can be then denoted by p_{ij}:

$$p_{ij}(t) = p_{ij} = P(X_t = x_i|X_{t-1} = X_j)$$

For example, the probability of a passage in s steps from a state s_i to generic state s_j for all the possible k states is given by the so-called *Chapmann-Kolmogorov equation*:

$$P_{i,j}^s = \sum_k P_{i,k}^r P_{k,j}^{s-r} \ : \ r \leq s \tag{3.2}$$

---

[2]For a comprehensive treatment see Aoki (2002, chap. 3), Landini (2005, chap. 6) and Kubo et al. (1978)

[3]Thus, analogously to Markov fields, that refer to a system of neighborhood of i, $N(i)$, Markov processes refer to a subset of temporal index $t \in \mathbf{T}$.

A jump Markov chain can remain in a state s_k for a certain time and then move to another state s_h. Define $T_{k,h}$ as the *waiting time* for the passage from state s_k to state s_h and the parameter $a_{k,h} = a(s_k, s_h)$ as the *transition rate* from state s_k to state s_h. For the properties of Markov processes, the random variable $T_{k,h}$ has the following exponential distribution:

$$P(T_{k,h} > \Delta) = \int_t^\infty a_{k,h} e^{-a_{k,h} u} du \tag{3.3}$$

with cumulative function:

$$F(t) = 1 - e^{-a_{k,h}\Delta} = P(T_{k,h} \le \Delta) \Rightarrow P(T_{k,h} > \Delta) = e^{-a_{k,h}\Delta} \tag{3.4}$$

Suppose that there is no *absorbing state*, so that the process cannot remain locked in any of the possible states and, therefore, it can move from state s_k to any s_h of the possible M states with probability $P_{k,h}$. The condition for normalization is:

$$\sum_{h=1}^M P_{k,h}(t) = P_{k,k} + \sum_{h\neq k}^M P_{k,h}(t) = 1 \Rightarrow \sum_{h\neq k}^M P_{k,h}(t) = 1 \tag{3.5}$$

Then, supposing that:

$$P_{k,h}(\Delta) = P(T_{k,h} \le \Delta) = 1 - e^{-a_{k,h}\Delta} \cong a_{k,h}\Delta \;:\; h \neq k$$
$$P_{k,k}(\Delta) = P(T_{k,k} > \Delta) = e^{-\lambda_k \Delta} \cong 1 - \lambda_k \Delta$$

equation (3.5) becomes:

$$\sum_{h=1}^M P_{k,h}(\Delta) = (1 - \lambda_k\Delta) + \sum_{h\neq k}^M (a_{k,h}\Delta) = 1$$

that, with opportune modifications, can be written as (Landini, 2005):

$$\lambda_k = \sum_{h\neq k}^M a_{k,h} \tag{3.6}$$

The sum λ_k is defined as the *jump rate* of state s_k and it is determinated by the sum of all the transition rates from s_k to all the other states. In a generic jump markovian process, for all i and j, the Chapmann-Kolmogorov equation (3.2) becomes:

$$P_{i,j}(t+s) = P\{X_{t+s} = s_j | X_s = s_i\} = \sum_k P_{i,k}(t)P_{k,j}(s)$$

and then:

$$P_{i,j}(t + \Delta) - P_{i,j}(t) = \left[\sum_{k \neq j} P_{i,k}(t)P_{k,j}(\Delta)\right] - [1 - P_{j,j}(\Delta)]P_{i,j}(t)$$

Finally, in order to evaluate the variation of probability in an infinitesimal time interval (and then for $\Delta \to 0$), using the result obtained until now and taking the limit of incremental ratio, it is possible to get:

$$\frac{dP_{i,j}(t)}{dt} = \sum_k a_{k,j}P_{i,k}(t)$$

where $a = \left\{\sum_{k \neq j} \lim_{\Delta \to 0} \frac{P_{k,j}(\Delta)}{\Delta}\right\}$.

This is called *forward Kolmogorov equation*, if it is formulated with respect to the final state, as in the above example, or *backward Kolmogorov equation* if it is respect to the starting state. In the following, the master equation will be specified as a forward Chapman-Kolmogorov equation. More precisely, master equation is a kind of Chapman-Kolmogorov equation that is representable as a first order differential equation of balance among in fluxes into a state y and out fluxes from state y. It can be expressed in the following way:

$$\begin{aligned}\frac{dP(y,t)}{dt} &= \sum_{x \neq y} P(x,t)a(y|x,t) - \sum_{x \neq y} P(y,t)a(x|y,t) = \\ &= \sum_{x \neq y}[P(x,t)a(y|x,t) - P(y,t)a(x|y,t)]\end{aligned} \qquad (3.7)$$

The equilibrium among in fluxes and out fluxes is then given by:

$$\sum_{x \neq y} P(x,t)a(y|x,t) = \sum_{x \neq y} P(y,t)a(x|y,t) \qquad (3.8)$$

that is the so-called *Kolmogorov condition*. If it holds for all pairs of the possible states in the system, it is defined as *detailed balance condition*.

Master equation is then revealed as a probabilistic structure by which it is possible to fully model a dynamic stochastic system and to describe its behaviour, embodying all known information. In particular it is possible to verify if a steady state solution exists and which is the equilibrium distribution. The main drawback is that, as demonstrated in Risken (1989), an analytical solution for master equations can be obtained only under very specific conditions. Aoki, in his books, identifies two main approximation methods to bypass this limit and then to find a stationary solution: the Van

Kampen's (van Kampen, 1965) and the Kubo's methods (Kubo et al., 1978)[4]. In this work (section 6.2), I shall apply a third method, proposed in Aoki (2002), that, using lead and lag operators, permits to obtain an approximate solution by means of Taylor series expansion of the master equation, once it has been conveniently transformed. In all cases, in order to exploit one of the cited procedures, the form of master equation has to be modified, and, specifically, it has to be split in two components: the *drift*, i. e. the tendency value of the mean of the state variable, and the *spread*, that quantifies what Aoki and Yoshikawa (2006) define as micro-fluctuations, i. e. the aggregate variations around the tendency value.

3.3　Hammersley and Clifford theorem: a functional solution for the master equation

Hammersley and Clifford demonstrate, in a theorem that circulated for many years without being published, that, under opportune conditions, for each Markov random field there exists one and only one Gibbs random field, defining also the functional form for the conjunct probability structure once the neighborhood relationships have been identified[5]. Brook's lemma (Brook, 1964) defines local characteristic of Markov chain, taking into account all the possible macro-states (i.e. the possible number of occupancies of each state, from 0 to X) of the system and states the condition under which the process is stationary. Therefore, using a markovian model we can approximate stochastic interaction among adjacent agents using Gibbs formalization, i. e. the potential. If detailed balance condition holds, condition (3.8) can be rewritten in the following way:

$$P^e(x)a(y|x) = P^e(y)a(x|y) \Rightarrow P^e(y) = P^e(x)\frac{a(y|x)}{a(x|y)} \quad \forall\,(x,y)$$

And then:

$$P^e(x_1) = P^e(x_0)\frac{a(x_1|x_0)}{a(x_0|x_1)} \;,\;\; P^e(x_2) = P^e(x_1)\frac{a(x_2|x_1)}{a(x_1|x_2)}$$

and so on. Thus, according to Brook's lemma, in a number of states equal to k the equilibrium probability will be given by:

$$P^e(x_k) = P^e(x_0)\prod_{h=0}^{k-1}\frac{a(x_{h+1}|x_h)}{a(x_h|x_{h+1})} \tag{3.9}$$

---

[4]For a detailed exposition see the usual Aoki (2002) or Landini (2005).
[5]See Hammersley and Clifford (1971) as demonstrated in Clifford (1990).

Finally, being the markovian process a markov random field and due to the Hammersley and Clifford theorem, its probability distribution is Gibbs. Then, provided that detailed balance condition holds, we have:

$$P^e(x) = Z^{-1} exp(-U(x)) \tag{3.10}$$

where U is the known Gibbs potential. Equation (3.10) represents the stationary probability for state x and then for the master equation (3.7).

3.4 Entropy measures and MaxEnt problem

Following Balian (1991), statistical entropy means a measure of uncertainty or disorder associated with a probability distribution. It can be referred to any system to evaluate the uncertainty among its alternative events or configurations. Formally, $p_h = p(\omega_h)$ expresses the probability for the h-th outcome to happen. Let us define ω_h as a configuration within a given state space Ω and p_h as the probability access to such a state for microscopic agents in the system. Then, consider a system $\mathcal{S}_N$ of N elementary constituents, each occupying one of the H configurations in Ω, and $0 \le n_h \le N$, i. e. the occupation number of the h-th configuration so that $p_h = n_h/N$ turns out to be the probability access. If each configuration is equally probable, $p_h = H^{-1}$ leads to the maximum level of uncertainty while, on the contrary, if $\exists! \ \omega_r \in \Omega : p_r = 1$ indicates the minimum level of uncertainty since all other configurations have zero probability access.

A macroscopic state for a system is given by a set of admissible configurations with respect to some macroscopic constraints: when a configuration is admissible, it can then be considered consistent with the macroscopic constraints. The uncertainty associated to a given macroscopic state of the system is defined by the the vector of occupation numbers $\mathbf{n} = (n_1, \dots, n_h, \dots, n_H)$. Now, indicating with $\mathbf{n}_0 = (n_1 = N/H, \dots, n_h = N/H, \dots, n_H = N/H)$ the maximum uncertainty vector of occupation numbers and with $\mathbf{n}_m = (n_1 = 0, \dots, n_m = N, \dots, n_H = 0)$ one of the admissible zero uncertainty situations, it is verifiable that there exists a great number of intermediate situations, growing with the number of configurations. The difference in probability between two macro states is evaluable in terms of entropy. Since a macroscopic state can be considered as a disposition of events over a state space, its uncertainty level is given by the quantity of information each single event carries on to the definition of the macro state.

Consider the state space as a set of configuration into which the N constituents of the system can be partitioned. In other words, the volume of N constituents can be classified into H classes. The number H of states is an indicator of variety for the system and the entropy measure is an index of variety. Let us consider the probability p_h as the weight of the *importance* of the h-th state, with respect to the others, given by some function of the kind $p_h = u_h / \sum_{h \leq H} u_h$ where u_h is a function measuring the *importance* of the each state. As Straathof (2003) pointed out, an efficient variety index $\mathcal{V}$ has to satisfy three main properties:

1. If $p_h = H^{-1}$ $\forall h$ then the index $\mathcal{V}$ is monotonically increasing in H;

2. $\mathcal{V}(\Omega) = \mathcal{V}(\Gamma) + \sum_j p_j \mathcal{V}(\Omega^j)$ where $\Omega^j \subset \Omega$ such that $\bigcup_j \Omega^j = \Omega$ with $\Omega^p \cap \Omega^q = \oslash$ $\forall p \neq q$ and Γ is the collection of all possible subset of Ω;

3. $\mathcal{V}$ is continuous in all p_h.

Property 2 states that global variety is the variety of all subsets plus the average value of all the subsets. This implies that the total entropy at a high level of aggregation is lower than the one at a more detailed level of aggregation. As a result, setting up a state space for macroscopic quantities and applying the same structure to microscopic quantities, the aggregate entropy is lower than the microscopic one. Property 3 states that an infinitesimal change in the importance of a state makes a marginal jump for the global entropy; then, the marginal gain or loss of importance of a state does not affect the global uncertainty level for the system.

The only index which satisfies these three properties is the *Shannon entropy measure* that is defined (Straathof, 2003) as:

$$S(\Omega) = -H \sum_{\omega_h \in \Omega} p(\omega_h) \log p(\omega_h) : p(\omega_h) = \frac{u(\omega_h)}{\sum_{\omega_h \in \Omega} u(\omega_h)} \quad , H = |\Omega| \quad (3.11)$$

where $u : \Omega \to \mathbb{R}_+$ is the importance function.

For our purposes, I exploit and utilize here the entropy maximization inference that has been developed and improved in the works of E. T. Jaynes, to which I refer the interested reader[6]. The maximization of the entropy functional (Jaynes, 1957), known in the specialized literature with the acronym of MaxEnt, ends up with the maximum likelihood estimation of probability distribution function in Gibbs form. This estimation will emerge as the less biased, the most probable, and most smoothed. Some other works have already presented applications of MaxEnt in economics[7]. Our goal is to estimate the distribution function for a variable with little existing information,

---

[6]Namely Jaynes (1957, 1968, 1979) among others.

[7]See for instance Foster (2004); Liossatos (2004).

improving the estimation when other informations become available, embodying them as constraints in an optimization problem. The argument of the optimization problem is the statistical entropy of the system, measured by the Shannon entropy as done in Aoki's works[8].

The justification of its adoption consists basically in its relative simplicity in obtaining Maximum Likelihood estimations in general conditions (Grendar and Grendar, 2002) and in its wide utilization in hard and social sciences to solve this kind of problem. Conceptually, MaxEnt may be regarded as an application of the Laplace principle of insufficient reason, according to which, with little or no information, it is opportune to assign to all the possible events the same probability, i. e. to assume a uniform distribution. Jeynes stated that the distribution function that maximizes the entropy has to be preferred since distributions that display a low level of entropy return the lowest level of fitting when applied to data. This property has been demonstrated in asymptotic conditions by van Campenhout and Cover (1981). To coin their words: *"the probability distribution which maximizes the entropy is numerically identical with the frequency distribution which can be realized in the greatest number of ways..."* (Jaynes, 1968) *"...thus associating maximum entropy with a definite frequency (or maximum likelihood) interpretation"* (van Campenhout and Cover, 1981).

In statistical mechanics, the entropy of a macro-state of a system is nothing but the average logarithm of the probability of occupation numbers for configurations of a state space or, in other words, the logarithm of the number of micro-states specified in terms of the state variables, i. e. the number of ways in which a macro-state can be realized.

Let us suppose to have a system S_N of N agents, e. g. firms, and that each firm produces output at a rate y_i; classifying each of them into M possible states of the kind (time references are suppressed):

$$\omega_k = \{y_i \in [a_k, b_k)\} \quad : \quad n_k = |\omega_k| \tag{3.12}$$

A single firm can occupy only one state at time, thus n_k measures the occupation number for the state ω_k. It follows that, specifying a theoretical model of the kind: $y_k = F(\mathbf{x}(\omega_k))$ for the control variable, as a function of a vector of state variables, then:

$$\sum_{\omega_k \in \Omega} n_k = N \quad and \quad \sum_{\omega_k \in \Omega} n_k y_k = Y \tag{3.13}$$

---

[8]See in particular the initial chapters in Aoki (1996).

Since we are dealing with two macroscopic variables N and Y, the entropy of the system can be indicated as:

$$S(N,Y) = \log W(N,Y) \tag{3.14}$$

where $W(N,Y)$, here, is the so-called multiplicity factor of the macroscopic state. It counts the number of ways leading the system to a certain macro configuration. Supposing that, for each ω_k, the n_k firms can assume m_k internal different micro-states, according to Aoki (1996), we can compute:

$$W(N,Y) = \sum_{n_k} \prod_k \frac{(n_k + m_k - 1)!}{n_k!(m_k - 1)!} \tag{3.15}$$

As a consequence, the basic MaxEnt problem is given by:

$$\begin{cases} \max S(N,Y) = \log W(N,Y) \, s.t. \\ \sum_{\omega_k \in \Omega} n_k = N \\ \sum_{\omega_k \in \Omega} n_k y_k = Y \end{cases} \tag{3.16}$$

As Aoki (1996) demonstrates, the multiplicity number is too difficult to be computed and so one can maximize the entropy finding the maximum for the biggest term under the two constraints in order to get the following approximated solution:

$$n_k^* = \frac{m_k}{e^{\gamma_k} - 1} \quad : \quad \gamma_k = \alpha + \beta y_k \tag{3.17}$$

where α and β are Lagrange multipliers. Then the entropy for the system at the macroscopic state is:

$$S(N,Y) \cong \alpha N + \beta Y - \sum_k (m_k - 1) \log(1 - e^{-\gamma_k}) \tag{3.18}$$

We define now the following partition function, obtained as the Laplace transform of multiplicity factor:

$$Z(\alpha, \beta) := \sum_{N=0}^{\infty} W(N,Y) e^{-\alpha N - \beta Y} = \prod_k \sum_{n_k} \frac{(n_k + m_k - 1)!}{n_k!(m_k - 1)!} e^{-n_k(\alpha + \beta y_k)} \tag{3.19}$$

It is possible to demonstrate (Landini, 2005) that the Lagrange multipliers in equation (3.18) act as semielasticities of the partition function. Indeed:

$$-\frac{\partial \log Z}{\partial \alpha} = \langle N \rangle \quad and \quad -\frac{\partial \log Z}{\partial \beta} = \langle Y \rangle \tag{3.20}$$

Then, the partial derivatives of the cumulant generating function gives back the expected value for the volume of agents and their output.

In section 6.1, I make use of statistical entropy maximization to obtain statical probability functions for the Markov field modeled in the next part, obtaining, with a simplified procedure, a close solution of system (3.16).

Part II

A dynamic stochastic model for business fluctuations

Chapter 4

Modeling financial fragility

> *"I suspect that the unwillingness to speak of workers in recession*
> *as enjoying "leisure" is more a testimony to the force of*
> *Keynes insistence that unemployment is "involuntary" than a*
> *response to observed phenomena"*
> (Lucas, 1977).

> *Marge: "What happened to you Homer?*
> *And what have you done to the car?"*
> *Homer: "Nothing."*
> *M.: "I don't think it had broken axles before."*
> *H.: "Before, before! You're living in the past, Marge!*
> *Quit living in the past!"*
> (From an episode of "The Simpsons").

This part introduces, develops and solves a model of financial fragility defined in a stochastic environment. As anticipated, business cycles are modeled in a New Keynesian perspective, and precisely following the bankruptcy approach introduced in Greenwald and Stiglitz (1990, 1993), as re-elaborated by Delli Gatti et al. (2005, 2007). Analytical and solution methods are taken from Aoki (1996, 2002) and Aoki and Yoshikawa (2006).

The original works of Greenwald and Stiglitz modeled a sequential economy in which identical isolated firms are subject to iid shocks in price and exposed to bankruptcy risk. In a market where firms are fully rationed on equity market, and there is uncertainty on demand side, the evolution on investment is influenced by financial structure of firms. Indeed, firms' optimization process reflects their risk-avoidance behavior, in the sense that their decisions (in terms of optimal production) take into account their level of net worth and their stock of liquid assets, in order to avoid bankruptcy.

Uncertainty and risk-aversion shift firms' supply curves to the left and this allows to explain three facts that puzzle traditional economists: first, the transmission and the persistence of small idiosyncratic shocks, since the shift of a firm supply curve determines a shift in the demand curves for other firms; second, why, even if a firm lowers its production, real wages do not rise; third, why sectors with flexible prices may record greater variability in production. In particular, the second paper (Greenwald and Stiglitz, 1993) succeeds in importing in a macroeconomic framework the micro level asymmetric information, and the connected problems on hazard behaviors and adverse selection, stressing the role of balance sheet variables in determining investments[1]. In particular, the major consequence of imperfect information is the restrictions for firms in raising funds on equity market. Empirical evidence shows that new equity issues represent a small fraction of capital raised and they are not used at all for financing working capital (Mayer, 1990)[2]. Moreover, Lintner (1971) shows that firms usually do not change the level of dividend paid, in order to avoid negative signals to investors[3] .

To briefly sum up the model's mechanism, firms plan their optimal production (and thus the level of investments) taking into account the expected costs of bankruptcy. These costs are weighted for the probability of failure, that is a function of their net worth. To finance investments, being fully rationed on equity market, firms recur to debt. When they sell output on the market, they face idiosyncratic price shocks. These shocks are stochastic, in order to reproduce the uncertainty on the demand side. Since price shock is a random variable, firms' profits and returns for lender are expressed as expected values. Greenwald and Stiglitz brilliantly demonstrate how uncertainty (and risk avoidance) determines the level of the equilibrium path of growth, and amplifies the effects and the persistence of small shocks, in an economy in which investments and credit supply are tightened proportionally to the expected bankruptcy costs. A degree of heterogeneity was implicit in this original construction[4].

As already noted (Gallegati, 2002), introducing financial hierarchy among the sources of funds (Myers and Majluf, 1984) in the original model of Greenwald and Stiglitz (where firms are supposed to be all identical, at least con-

---

[1]On this aspect see Fazzari et al. (1988).

[2]Formal models to explain these evidence have been proposed by Myers and Majluf (1984) and Greenwald et al. (1984).

[3]Greenwald and Stiglitz (1993) assumes that firms pay a fixed dividend. In the present model I simply assume that own capital is remunerated at the same level of borrowed capital, as in Delli Gatti et al. (2005).

[4]In general, heterogeneity is always implicit in a market where asymmetric information is present.

sidering their marginal source of financing), firms turn out to be different as regards strategy for financing investment and, eventually, financial structure. Gallegati (2002), first introducing heterogeneous firms in a Greenwald and Stiglitz approach, adopted a classification of firms that divides them, to use his terminology, between *liquidity constrained* and *demand constrained*, being the former firms with insufficient internal funds respect to the desired level of investment, and the latter firms with an excess of availability of own capital respect to their investment opportunities. The proportion of firms of the two types existing in the system at a given time follows a binomial stochastic process, which appears to be particularly suitable for such a dichotomic variable. The model is dynamic and it is analytically solved using implicit functional forms. The obtained solution permits to get indications about the absence of a natural equilibrium level of production (given that the percentage of demand and liquidity constrained firms was stochastically determined, the growth path cannot be unambiguously predetermined) and about the non-neutrality of money in the long-run.

In the model here proposed, the division among the two groups of firms is between firms that risk bankruptcy and firms that are, temporarily "safe"; it leads to two remarkable differences between the two modeled structures. First, this kind of classification comes out to be endogenously determined by firms' dynamics, and, second, permits to isolate the role and the effects on aggregate output of bankruptcy costs, weighted by failure probability, the latter being endogenously quantified in the model's mechanism. Indeed, stochastic dynamics, that in Gallegati (2002) was defined by hypothesis, here is completely endogenous, without formulating any assumption about firms' distribution. As will be shown in the following sections, statical inference is performed by means of statistical entropy maximization, while stochastic dynamics is determined by the probabilities of transition among states that are the results of, on the one hand, the optimizing process of the firms and, on the other, of their interaction.

Delli Gatti et al. (2005, 2007), introduce in the original framework two innovative elements, characterizing, in this way, economy as an evolving complex system: firms heterogeneity (in size and financial condition) and indirect interaction among them. Initial capital of firms is randomly drawn. Firm's optimization in terms of maximum profit determines the desired level of investment, and to reach it firms use, at first, their own capital and then recur to debt. Interaction is made effective through the bank (or the banking system), that acts as lender to firms. Bank accords credits in a quantity proportional to its equity and allots this total credit on the basis of firm's equity; the failures of firms reduce the equity of the bank, diminishing the availability of credit and making interest rates rise. This interaction, mediated by the bank and the cost of the credit, is at the root of business fluctuations. Then, as the model runs, firms become more and more different, displaying, as the transition period ends, a power law distribution of capital. The quantitative previsions of these models, obtained by means of computer simulations, replicate a number of empirical stylized facts, strengthening the idea that the economy would be better represented as a complex dynamical system rather than a mere sum of identical and perfectly informed agents.

According to Bak (1997) the occurrence of a power law is the consequence of the interaction among system units that, reacting to idiosyncratic shocks, leads to a complex critical state in which no attractive points or states emerge. This feature of the system may be reassumed under the notion of Self-Organized Criticality. Therefore, the emergence of such a probabilistic structure (also defined as "scale-free distribution") means that it is not possible to properly define a characteristic (or representative) scale for agents. Moreover, equilibrium exists only as asymptote, along which the system moves from one unstable (critical) point to another.

An approach of this kind appears to be more promising and more informative compared to one that assumes that the system is in equilibrium, whatever the equilibrium is, limiting the analysis to record Homer's car conditions, even though these are completely out of a desirable normality. Keynes (1936) remarks on the relevance of analysis of transitions, since it may reveal what is behind the surface of gross economic indicators and allows to enlighten the mechanisms of transmission from the micro level to the macro (and *viceversa*). Without a representation of these structures it is rather difficult to imagine an effective economic policy. In order to perform such an analysis we need proper tools and instruments. Let us start to build them.

Chapter 5

Hypothesis

In this chapter the assumptions at the basis of the model are detailed. In the first section the stochastic mechanism of the system is modeled. Then, in section 5.2, the hypothesis regarding firms and market are described.

5.1 Structure of the system and definition of states

In this section the probabilistic structure of the economic system is defined. Three things are worth being preliminarily stressed. First, there is a finite but large number of agents that interact. Second, in each unit of time, agents are heterogeneous with respect to some particular characteristics. They are grouped in different types, basing on these specified features. The composition of the groups changes over time since agents may modify their characteristics, switching from one particular type to another. Finally, it must be observed that analysis is not focused on the agent or on the micro level, but on their groups or types or, as properly said, state. A meso or macro state is consistent with different states at micro level, or, put in different words, it is not possible to infer the macro-results by the situation of the agents at micro level[1].

As in standard economic modeling approach, each of them solves an optimization problem and these solutions determine the evolution at macro level. But, differently from standard models, on the one hand, optimization problems are not unique and identical for all agents but are different (in the inputs and, therefore, in the solutions) and, on the other hand, optimization's solutions are the outcomes not only of the single agent's optimization problem but also of the interaction among them. As already mentioned, this

---

[1]This attribute of complex systems characterizes this class of models from traditional models with heterogeneous agents (Caballero and Engel, 1991).

interaction is represented as a mean-field interaction, by which the stochastic factors that determine the dynamics of the meso level (state) variables are subject to feedback effects from the macro-level. Aoki and Yoshikawa (2006) indicate two classes of methods to model this kind of problem: stochastic dynamics and random cluster formation. In this work I adopt the first method, which consists, basically, in two steps. First, dynamics of states is described by means of a master (or Chapman-Kolmogorov) equation. Second, an (approximate) solution of master equation is found in order to identify an (eventual) equilibrium distribution and to determine a functional form that describes the fluctuations around this equilibrium path.

This paragraph reports the hypothesis that are at the basis of this stochastic dynamics, and in particular the definition and the structure of states[2]. Our system is articulated in two micro states (that is to say two types of firms). This simplification is basically due to the fact that two states are sufficient to articulate a model that, consistently with the referenced New Keynesian literature, permits to isolate the effect of bankruptcy costs on one type of firms and, then, on aggregate dynamics. Moreover, the application of recently proposed methods for systems with higher order of states (de La Lama et al., 2006) would entail a price too high in terms of computational complication respect to the expected improvement in the realism of the model.

A single firm can be, at each time, in one of the two states, depending on its financial soundness, quantified by the equity ratio, i. e. the ratio among the net worth and the total assets. Therefore there can be two types of firms: the "good" firms, that have a high equity ratio, and the "bad" firms, that have a low equity ratio which exposes them to the risk of demise[3].

The other assumptions are the following:

- the system works in continuous time:

$$t \in \mathbb{T} \subset \mathbb{R}_+$$

This is due to both analytical and epistemological reasons. Continuous time, on the one hand, is consistent with the hypothesis that the stochastic evolution of the state variable follows a continuous time Markov chain, allowing to employ some analytical tools that could not be used in discrete time. Furthermore, it appears more suitable with

---

[2] As defined in section 3.2, a state is a set of information, sufficient to determine the evolution of probability distribution of system's configurations.

[3] To be precise, there is also a third type, the "ugly" firms, i.e. the failed ones, as explained below.

a complexity approach, getting rid of a very stimulating stream of research on econometric analysis of time series (Hinich et al., 2006). As
has been observed,

> "Most time series can be conceived as a continuous flow of
> data through time that is measured by some procedure. The
> measurement procedure used in engineering and science ap-
> plications is deliberative but in the social sciences the filtering
> and sampling methods that are used to generate discrete-
> time samples are rarely discussed explicitly. Each discrete-
> time observation $\tilde{x}(t_n)$ is treated as if it was the true value
> at time t_n rather than an average value around t_n"(Hinich
> et al., 2006).

In other words, they note that the recorded value of an observation
appears to be a function of the particular sampling frequency that the
researcher (or whoever collects the data) has adopted;

- economic system is meant to represent the whole economy of a country
  or a part of it, as e. g. a sector. It is populated by a fixed number of
  firms N, indexed by the subscript i, for any given time:

$$N = N(t) \; \forall t \in \mathbb{T}$$

- the system's vector of states $\underline{\omega}$ specifies the microeconomic state at a
  point in time; it is identified by the financial condition of firms, i. e.
  by their equity ratio:

  - $\omega = 1$: state 1: equity ratio $< \bar{a}$:
  - $\omega = 0$: state 0: equity ratio $\geq \bar{a}$

where $\bar{a}$ represents the threshold of equity ratio, that identifies firms
that are in a critical financial situation, and, therefore, for which the
probability of bankruptcy is bigger than 0. Identifying with $a_i(t)$ the
equity ratio of the ith firms:

$$\underline{\omega}(t) = \{\omega_i(t) = HV\left(a_i(t)\middle|\; \bar{a}\right) \; \forall i \; \leq N\}$$

with the Heaviside function HV defined as follows:

$$HV\left(a_i(t)|\bar{a}\right) = \begin{cases} 1 & \Longleftrightarrow & a_i(t) < \bar{a} \\ 0 & \Longleftrightarrow & a_i(t) \geq \bar{a} \end{cases}$$

For the analytical treatment, all the a_i for firms in state x and y are approximated by the two variables a^1 and a^0, respectively. In this way it is feasible to obtain the mean-field approximation of interactions among agents. More precisely, $a^j : j = 0, 1$ can be regarded as a statistic S of all the equity ratios for each state:

$$a^j = S\{a_1, ..., a_i, ..., a_N\} \ : \ HV(a_i|\bar{a}) = j$$

Section 5.2 below specifies the balance sheet variables' relations and, then, it will be possible to endogenize the threshold as the minimum value of the equity ratio needed to avoid bankruptcy in the following unit of time.

- the number of firms in a generic state j in a given time is indicated by $N^j : j = 0, 1$ and its fraction by small letter: $n^j = N^j/N$. In what follows, the apex refers to the state, while the firm's index is at subscript;

- for hypothesis, the dynamics of the occupation number N^j follows a suitable continuous time jump Markov process. Indeed, at each point in time, agents modify their targets in reaction to external shocks or to any modification in the arguments of their objective functions. This evolving optimization modifies their situation and possibly makes them pass from one state to another. These adjustments happen at each instant for an unknowable number of agents, inducing modifications in their environment. This process gives rise to multiplicative and feedback effects that are governed by a composite stochastic law (the probability of being hit by a certain shock, the probability of changing strategy or type in consequence of the shock, the probability of observing a subsequent change at macro level). Stochastic Markov processes integrate, at each time, a given starting situation and all the possible transitions to other situations and thus emerge as a particularly suitable representation for the modeled dynamic structure. It gets rid at the same time of the unpredictability at micro level and of the volatility at macro level. For each of the two states, N_k is defined as the number of firms occupying one of them in a given instant, with $0 \leq N_k \leq N$. Therefore, consider a state space $\Omega = (x, y)$ with a counting measure $N_k(t) : \Omega \times \mathbb{T} \to \mathbb{N}$, so that $(\Omega, N_{(.)}(t))$ can be regarded as a countable sample space. Then, indicating with x the case of firms with equity ratio below the threshold $\bar{a}$ and with y the alternative case, the formal notation is:

$$\omega = x \Leftrightarrow \omega_i(t) = 1 \ \lor \ \omega = y \Leftrightarrow \omega_i(t) = 0 \qquad (5.1)$$

so that the counting measure $N_{(.)}(t)$ evaluates the cardinality of microstates;

- the a-priori probability of $\omega = 1$ is indicated by η:

$$p(\omega = 1) = \eta \Leftrightarrow p(\omega = 0) = 1 - \eta$$

As a first step, this variable is considered as exogenous, specifying all the other features of the model without formulating any other hypothesis about this probability. In the following chapters, the stationary solution of the problem will allow to formulate an explicit endogenous functional formulation for it;

- firms entry into the system in state x. It appears reasonable to assume that new firms display a low equity ratio that put them at a bankruptcy risk[4];

- consequently to the above specified hypothesis a firm may fail (and thus exit from the system) only if it is in state x (i. e. if its probability of bankruptcy is bigger than 0);

- by assumption, in order to maintain constant the number of firms N, each bankrupted firm is immediately substituted by a new one. Therefore failures of firms do not modify the value of N^1.

As configured by these assumptions, our system describes four theoretical kinds of transition: 2 in entry in x (new firms and firms from state y) and 2 in exit from x (firms that fail and firms that improve and pass to state y). Since failures of firms do not modify the occupation numbers, the attention is focused on the two transitions from one state to another.

In order to model these flows, we have to properly define the probabilities of transition at agents' level (*transition probabilities*) and at aggregate level (*transition rates*). The former are the probability for a single agent to switch from one micro state to the other in a given instant. Their calculation is shown in the next section. For the moment it is enough to define:

- ζ as the probability of transition from state y to x (firms whose financial position has deteriorated from one period to another, with equity ratio that becomes lower than $\bar{a}$);

- ι as the probability of the inverse transition (firms whose equity ratio has improved becoming bigger than $\bar{a}$).

---

[4]This is also consistent with existing empirical evidence (Dunne and Hughes, 1994).

Moreover, I indicate with μ the probability of bankruptcy for a firm. The particular configuration assumed by agents, jointly determined by these probabilities, gives origin, as a consequence, to a particular macro-state at aggregate level. This macro state is identified by the occupation number. Then, the set of macro states is represented by all the possible occupation numbers N^j with $j = 0, 1$. Once a particular macro-state $N^j(t)$ is verified, it can be modified by a "birth" in state j, for a transition from the other state, or by a "death", i. e. a transition to the other state. Therefore, in order to describe the dynamic behavior of the system, we need to specify the aggregate probability to observe, in a unit of time, a "jump" of an agent from one state to another, and a consequent variation in the occupation numbers, given a starting macro state. This measure of probability is quantified by transition rates. Given that N does not change over time, it is enough to observe one of the states. Therefore, in order to simplify notation, we can take as state variable the numbers of firms in x, identifying N_k uniquely as the number of firms occupying it in a given instant.

Then, using the symbols adopted in Aoki (2002, page 42), in a unit of time we can observe a unitary increment or decrement in N_k, that is indicated, respectively, as r (a step to the right toward $k + 1$) and as l (a step to the left toward $k - 1$). At aggregate level, the probabilities of a "death" (d) and of a "birth" (b) obviously depend on N_k, since a large number of firms in x implies, *ceteris paribus*, a relatively high probability of exit from it and a relatively low probability of an entry in it. Formally:

$$b(N_k) = r(N_k + 1 | N_k) = \zeta \frac{N - N_k}{N}(1 - \eta)$$
$$d(N_k) = l(N_k - 1 | N_k) = \iota \frac{N_k}{N} \eta \tag{5.2}$$

Then, I specify transition rates, respectively, as λ for entries into state x, and as γ for exits out from the same state, according to the following formulation:

$$b(N_k) = \lambda \left(\frac{N - N_k}{N} \right) : \quad \lambda = \zeta(1 - \eta)$$
$$d(N_k) = \gamma \left(\frac{N_k}{N} \right) : \qquad \gamma = \iota \eta \tag{5.3}$$

Thus, transition rates are given by the probability for a firm to move from one state to another, pondered by the probability of being in the starting state. Multiplying the result by the actual occupation numbers, we obtain the probability of observing such a transition in a unit of time. Transition rates λ and γ are then the key variables and, in what follows, we will focus our attention on them in order to describe the stochastic dynamics of our system. In statistical mechanics terms, this kind of system can be defined as a statistical ensemble with conservative cardinality, described by a continuous time Markov process over a discrete state space with the structure of a birth-death process. A scheme of the system's mechanism is depicted in figure 5.1, that shows how transitions between states (indicated as x and y) are regulated by transition rates λ and γ. The rate of exit from the system is $\mu\eta$, that, given the hypothesis, represents also the rate of entry into the system.

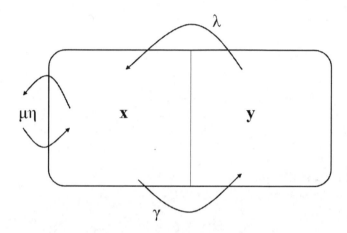

Figure 5.1 – Structure of the system.

5.2   The Firms

Once the probabilistic structure of the system, its states and the stochastic laws that govern its evolution have been defined, in what follows the hypothesis about agents' economic behavior are specified. The assumptions regarding firms are adapted from the original models of Greenwald and Stiglitz (1990, 1993) as modified by Delli Gatti et al. (2005). The adjustments are mainly due to the opportunity to keep the computational mechanism as simple as possible in order to stress the aggregation problem and the proposed solution. At the same time, the framework maintains coherence with the original bankruptcy cost approach and a degree of heterogeneity among firms that is suitable to the stochastic dynamics detailed in the previous section.

Compared to the theoretical construction proposed by Delli Gatti et al. (2005) this model does not consider any credit market. That mechanism is replicated here without the necessity of modeling a banking sector. But, also in this model, firms' dynamics is determined by their interaction, since the

probability to observe a deterioration in their financial position is influenced
by the general financial situation of the economy, determinated by the number of firms in critical debt conditions. And, again in analogy with Delli Gatti
et al. (2005), interaction is not direct, being, in this work, a mean-field interaction, that is to say, mediated at a meso-level of aggregation. As diffusely
explained above, the dynamics of the model is determined not by the single
agents but by the evolution of the states and of their occupation numbers,
ignoring the changes intervening at the micro level.

5.2.1   Assumptions about firms behavior and market

At any instant, the supply side of the economy consists of finitely many
competitive firms indexed with $i = 1, ..., N$, each one located on an island.
The total number of firms N does not change over time. The market in which
firms operate is fully supply-determined, in the sense that they can sell all
the output they (optimally) decide to produce[5] . Firms are identical within
each state, as indicated above by $j = 0, 1$. Firms employ only one factor
in the production process, indicated as the physical capital K. The capital
stock never depreciates. The production function of a generic firm i is:

$$q_i(t) = 2(K_i(t))^{1/2} \tag{5.4}$$

where q is the physical output. Since it is perishable, all the production is
sold in the successive unit of time and so there are no stocks. It follows that,
at each time, the demand function for capital of a single firm will be equal
to:

$$K_i(t) = \frac{1}{2}(q_i(t))^2 \tag{5.5}$$

Basing on equation (5.5), firms adjust their capital stock in order to match
their demand, for hypothesis, without any adjustment cost. As a result,
firms invest an amount equal to $I_i(t)$ to reach that desired level of capital. It
is straightforward to note that $I_i(t)$ can be either bigger or smaller than 0,

---

[5]This assumption is consistent both with an equilibrium and with a disequilibrium *scenario*. In the *equilibrium scenario*, aggregate demand accommodates supply: households
and firms absorb all the output produced by the latter and the goods market is always in
equilibrium. In this scenario, aggregate investment must be equal to the sum of retained
profits and households' saving. Both investment and retained profit are determined in the
model, so that we have to assume that households' saving adjusts in order to fill the gap
between them. In the *disequilibrium scenario*, i.e. aggregate demand does not accommo-
date supply, so that the goods market is generally not in equilibrium. In this case, the
difference between aggregate investment on the one hand, and the sum of retained profit
and households' saving on the other must be assumed to take the form of involuntary
inventories decumulation.

since a firm that has to reduce its production has to also diminish its physical stock[6]. As in the referenced models, there exists a lag of one unit of time between firms' decisions and output sell.

The balance sheet identity implies that firms can finance their capital stock by recurring either to net worth, $A_i(t)$ or to bank loans, $B_i(t)$, so that:

$$K_i(t) = A_i(t) + B_i(t)$$

Under the assumption that firms and banks sign long-term contractual relationships (with no pre-fixed date of repayment), at each t debt commitments in real terms for the ith firm are $rB_i(t)$, where r is the real interest rate. Supposing that the market of capital is always in equilibrium (so that the real return on net worth equals the interest rate r), each firm incurs in financing costs equal to:

$$r(B_i(t) + A_i(t)) = rK_i(t) \qquad (5.6)$$

In Greenwald and Stiglitz (1993), the representative firm always recurs to debt to finance its investments, since costs are always bigger than equities. Here new investments can be either bigger or smaller than own capital (or even negative). According to the financial hierarchy approach (and to empirical evidence as shown in Fazzari et al., 1988), firms use own capital first and, if it is not enough to finance capital investment, recur to loans. The variation in the stock of debt in a period will then be equal to:

$$\delta B_i(t') = I_i(t') - A_i(t) \qquad (5.7)$$

with $t' - t = \delta t \to 0^+$. Analogously with Greenwald and Stiglitz (1990, 1993), firms are assumed to be fully rationed on the equity market, so that the only external source of finance at their disposal is credit.

In a context of asymmetric information, the rise of interest rate does not represent an efficient mechanism of credit rationing since it may determine adverse selection phenomena. Therefore, in this model, interest rate is assumed to be constant over time and equal for all firms, i. e. $r_i(t) = r \, \forall t, i$. The system itself provides a mechanism of credit rationing, given that demand of credit is endogenously limited by the model and cannot grow indefinitely, as demonstrated below.

The demand for goods in each island is affected by an iid idiosyncratic real shock. Since arbitrage opportunities across islands are imperfect, the individual selling price in the ith island is the random outcome of a market

---

[6] Consequently, negative investment implies negative variation of debt, i. e. disinvestment is used to pay debt.

process around the average market price of output $P(t)$, according to the law:

$$P_i(t) = \tilde{u}_i(t)P(t) \tag{5.8}$$

The random variable $\tilde{u}_i(t)$ has uniform distribution with $E(\tilde{u}) = 1$. Without loss of generality its support can be fixed in the interval $[0.75; 1.25]$. The choice of the range for $\tilde{u}$ does not affect probabilities, given the the normalization procedures detailed below. As in the original models these idiosyncratic and unanticipated micro shocks are at the roots of aggregate fluctuations, for the interaction among firms that multiplies their effects. Price shocks and interest rate represent the only exogenous input in the model, being all the other variables determined by the internal mechanism of the system.

Once a firm fails, it faces bankruptcy costs $C_i(t)$ growing with the size of firm (expressed in terms of sales), and quantified by:

$$C_i(t) = c(P_i(t)q_i(t))^2 = c(P(t)u_i(t)q_i(t))^2 \tag{5.9}$$

with $0 < c \leq 1$. Analogous functional forms are adopted also in Hennessy and Whited (2007) and Welch et al. (2003). Greenwald and Stiglitz (1993) use a linear function but their results, as they state, hold for any function convex in q. The justifications to introduce bankruptcy costs are identified by Greenwald and Stiglitz (1990) in the behaviour of managers, who act to avoid failure for two main reasons. First, given the impossibility of addressing the cause of default to bad luck or to managers' incompetence, bankruptcy may generate a stigmatization of managers' behaviour and, as a consequence, a reduction in their future earnings. Second, the imposition of a penalty for failure constitutes an incentive, since managers' contracts cannot state a participation in losses. As a further consequence, bankruptcy costs grow with size, since the number of employed managers is determined by the scale of operation.

Given all the above specified hypothesis, the profit function for a generic firm i can be expressed by:

$$\pi_i(t) = P(t)\tilde{u}_i(t)q_i(t) - r_i(t)K_i(t) - C_i(t)\mu(t) \tag{5.10}$$

where $\mu(t)$ is the probability of default introduced in section 5.1. The law of motion of equity for each firm is then:

$$A_i(t') = \pi_i(t) + A_i(t)$$

Transition probabilities

A firm goes bankrupted if it "consumes" all its own capital, thus if A_i becomes ≤ 0. It follows that bankruptcy happens when:

$$P(t)\tilde{u}_i(t')q^1(t) + A^1(t) - r(A^1(t) + B^1(t)) \leq 0$$

where the apex indicates the state. Given that $K(t) = A(t) + B(t)$ and that the only unknown variable in the above equation is the price shock $\tilde{u}_i(t')$, the bankruptcy condition can be opportunely expressed as a function of it:

$$\tilde{u}_i(t') \leq \left[\frac{P(t)}{P(t')}\right] \left[rK^1(t)/q^1(t) - a^1(t)\frac{K^1(t)}{P(t')q^1(t)}\right] \equiv \bar{u}(t')$$

Substituting equation (5.5) into the above expression and, without loss of generality, normalizing reference price $P(t) = P(t')$ to 1, it is possible to simplify the r.h.s. of the above equation in the following way:

$$\bar{u}(t') \equiv \frac{q^1(t)}{2}\left[r - a^1(t)\right] \tag{5.11}$$

Recalling that the random variable $\tilde{u}$ has support $[0.75; 1.25]$, the critical thresholds of price shock for having bankruptcy will be:

$$\begin{cases} \bar{u}(t) = 0.75 & if \ \bar{u} < 0.75 \\ \bar{u}(t) \leq \tilde{u} & if \ 0.75 \leq \bar{u} \leq 1.25 \\ \bar{u}(t) = 1.25 & if \ \bar{u} > 1.25 \end{cases} \tag{5.12}$$

Then, it is possible to indicate the probability of failure for a firm, $\mu(t)$, as the distribution function of $\tilde{u}(t)$:

$$\mu(t) = F(\tilde{u}(t)) = p(\tilde{u}_i(t) \leq \bar{u}(t)) = \frac{\bar{u}(t) - 0.75}{0.5} = 2\bar{u}(t) - 1.5 \tag{5.13}$$

Equations (5.11) and (5.13) permit to determine endogenously the threshold $\bar{a}$: indeed it can be interpreted as the minimum value of equity ratio which ensures the firm's survival, i. e. for which the probability of bankruptcy is equal to zero. Therefore, it can be expressed as:

$$\bar{a}(t') = r - \frac{1.5}{q^1(t)} \tag{5.14}$$

Formally, the threshold $\bar{a}$ is subject to changes over time, since it depends on the targeted optimal level of production for firms in state x in the previous unit of time. In what follows, for the sake of simplicity, I suppress the temporal indexes, indicating the threshold simply with $\bar{a}$. With a procedure analogous to the one used for determine $\mu(t)$, we can specify the transition probabilities ζ and ι as dependent variables of the price shock $\tilde{u}_i(t)$. Indicating the critical values, respectively with $\bar{u}^\zeta(t)$ and $\bar{u}^\iota(t)$, one obtains:

$$\tilde{u}_i(t) \leq \frac{q^0(t)}{2}(r + \bar{a} - a^0(t)) \equiv \bar{u}^\zeta(t)$$
$$\tilde{u}_i(t) > \frac{q^1(t)}{2}(r + \bar{a} - a^1(t)) \equiv \bar{u}^\iota(t)$$

since the thresholds, that for the (5.13) is equal to 0, here become equal to, respectively $(\bar{a} - a^0(t))$ and $(\bar{a} - a^1(t))$. The range of variation of the two thresholds is truncated as in (5.12), thus:

$$\begin{cases} \bar{u}^\varsigma(t) = 0.75 & if \ \bar{u}^\varsigma(t) < 0.75 \\ \bar{u}^\varsigma(t) \leq \tilde{u} & if \ 0.75 \leq \bar{u}^\varsigma(t) \leq 1.25 \\ \bar{u}^\varsigma(t) = 1.25 & if \ \bar{u}^\varsigma(t) > 1.25 \end{cases} \qquad (5.15)$$

and

$$\begin{cases} \bar{u}^\iota(t) = 0.75 & if \ \bar{u}^\iota(t) < 0.75 \\ \bar{u}^\iota(t) \leq \tilde{u} & if \ 0.75 \leq \bar{u}^\iota(t) \leq 1.25 \\ \bar{u}^\iota(t) = 1.25 & if \ \bar{u}^\iota(t) > 1.25 \end{cases} \qquad (5.16)$$

Once the thresholds have been quantified, it is straightforward to get the transition probability for each state:

$$\zeta(t) = p(\tilde{u}_i(t) \leq \bar{u}^\varsigma(t)) = 2\bar{u}^\varsigma(t) - 1.5 \qquad (5.17)$$

$$\iota(t) = 1 - p(\tilde{u}_i(t) \leq \bar{u}^\iota(t)) = -2\bar{u}^\iota(t) + 2.5 \qquad (5.18)$$

5.2.2   Firms objective function

A firm decides the optimal quantity to produce in order to maximize its profit, using the information at its disposal. Under the stated hypothesis the objective function of a generic firm i can then be expressed as:

$$\max_{q_i(t)} F(q_i(t)) := \mathbb{E}\left[P(t)u_i(t')q_i(t) - rK_i(t) - C_i(t)\mu(t')\right] \qquad (5.19)$$

By assumption, firms take into consideration the present level of failure probability, therefore $\mathbb{E}[\mu(t')] = \mu(t)$. Suppose that $\mathbb{E}[P(t')] = P(t') = P(t) = 1$ without loss of generality. Using equation (5.5) and considering that $\mathbb{E}[\tilde{u}] = 1$, the argument of the (5.19) can be rewritten as:

$$q_i(t) - rK_i(t) - C_i(t)\mu(t) = q_i(t) - r\frac{1}{2}(q_i(t))^2 - c(q_i(t))^2\mu(t)$$

The first order condition is then:

$$1 - rq_i(t) - 2cq_i(t)\mu(t) = 0$$

Consequently, there are two different optimal levels of production for firms in state x and for firms in state y, respectively:

$$q^{1*} = (r + 2c\mu(t))^{-1} \qquad (5.20)$$

$$q^{0*} = r^{-1}$$

since $\mu = 0$ for firms in state y. Therefore the aggregate production:

$$Y(t) = N^0(t)q^{0*} + N^1(t)q^{1*}$$

comes out to be dependent, *ceteris paribus*, on the occupation numbers N^0 and N^1. In the following chapters, through the the analysis of the evolution of occupation number's probabilities, it will be possible to describe the dynamics of output. In section 7.1, the aggregate supply function is reformulated in terms of the stochastic variables introduced in this chapter.

Finally, it is possible to quantify the upper bound of demand of money. Given the optimal levels of capital for each cluster of firms K^1 and K^0, the quantity of demanded credit reaches its maximum when a^1 and a^0 reach their minimum. As shown by equations (5.11) and (5.13), a^1 cannot be lower than $r - 2.5/q^1$, value for which the probability of bankruptcy becomes equal to 1, while the minimum level for a^0 is, by definition, $\bar{a} = r - 1.5/q^1$. For these values we have:

$$\lim_{a^0 \to \bar{a}} B^0 = \frac{(q^0)^2}{2}\left(1 - r + 1.5/q^1\right)$$

$$\lim_{a^1 \to min(a^1)} B^1 = \frac{(q^1)^2}{2}\left(1 - r + 2.5/q^1\right)$$

Consequently, indicating with $B = B^0 + B^1$ the total debt, the maximum level of demand of debt is:

$$max(B) = N^0\left[\frac{(q^0)^2}{2}\left(1 - r + \frac{1.5}{q^1}\right)\right] + N^1\left[\frac{(q^1)^2}{2}\left(1 - r + \frac{2.5}{q^1}\right)\right] \quad (5.21)$$

that cannot grow indefinitely since $q^0 = 1/r$ and $q^1 < q^0$.

Chapter 6

Stochastic inference

This chapter presents the probabilistic analysis of the above modeled economic system, facing, as a first step, the problem of how to estimate the probability for a firm to be in one of the two states in a given instant. In section 6.2 by means of the master equation, it will be possible to describe the stochastic evolution of the system and to eventually find an equilibrium solution.

6.1 Statical inference

6.1.1 The MaxEnt problem

The statical problem faced in this section permits to get a first insight into the system's mechanism. In particular, imposing probability to be consistent with macro-economic constraints (namely, linking it to the aggregate volume of production), it will be feasible to quantify a first, statical effect of the interaction. Coherently with the approach adopted in all this work, no hypothesis about the probability function is imposed and the problem is faced as no information is available[1]. In this context the use of statistical entropy emerges as a particularly useful instrument to make inference. The meanings and the method of solution of maximum entropy problems have been presented in section 3.4. Here the method is applied, returning the quantification of two indicators that will emerge as particularly important in dynamic analysis:

---

[1]Actually, as shown in section 9.2, one could infer the empirical structure of such an economy by studying the distribution of equity ratio across firms in a real economic system. Anyway the structure here presented is meant to work with no hypothesis about the features of the system or about its population.

- the probability for a firm to access one of the two states in an instant, given the macroeconomic conditions. Indeed, the probability for a firm to be among the "bad" or among the "good" largely depends upon the general condition of economy. This factor is embodied into the MaxEnt problem as a constraint;

- an index that, depending on the proportion of firms in the two states, comes out to be a synthetic indicator of the uncertainty in the system and, consequently, of how economy is performing, given that a bigger fraction of firms in state x means a lower level of aggregate production Y. This index will determine also the dynamics of probabilities, as detailed in section 6.2.

As anticipated in section 3.4, the solution of MaxEnt problem returns maximum likelihood estimation of probability distribution function in Gibbs form. This estimation will emerge as the less biased, the most probable, and the most smoothed. For its properties, that have been detailed in the same section, the adopted measure of entropy is the Shannon entropy measure.

6.1.2   Solution and estimations

In the system treated here, the problem has the following expression:

$$\max_{N^1, N^0} H(N^1, N^0) = -N^1 log(N^1) - N^0 log(N^0) \tag{6.1}$$

s.t.

$$\begin{cases} N^1 + N^0 = N \\ N^1 y^1(t) + N^0 y^0(t) = Y(t) \end{cases} \tag{6.2}$$

The first of the two equations in (6.2) ensures the normalization of the probability function, maintaining the number of firms in each group less or equal to the total number of firms in the system. In the second constraint, y^1 and y^0 represent the value of production of a single firm for both states. This condition ensures that all the wealth in the system is generated by firms in the two states. Therefore, it introduces interaction among agents, linking the estimation to the business cycles and, in this way, to the behavior of the other firms in the system[2]. The lagrangean is:

$$\begin{aligned} \ell = \ & -N^1 log(N^1) - N^0 log(N^0) + \delta_1(t)N^1 + \delta_1(t)N^0 - \delta_1(t)N + \\ & +\delta_2(t)N^1 y^1(t) + \delta_2(t)N^0 y^0(t) - \delta_2(t)Y(t) \end{aligned}$$

---

[2]Maintaining a normalized reference price $P(t) = 1$, we can use both quantities or values to express production levels.

with first order conditions[3]:

$$
\begin{cases}
\frac{\partial \ell}{\partial N^1} = -log(N^1) - 1 + \delta_1(t) + \delta_2(t)y^1(t) \\
\frac{\partial \ell}{\partial N^0} = -log(N^0) - 1 + \delta_1(t) + \delta_2(t)y^0(t) \\
\frac{\partial \ell}{\partial \delta_1(t)} = N - N^1 - N^0 \\
\frac{\partial \ell}{\partial \delta_2(t)} = Y - N^1 y^1(t) - N^0 y^0(t)
\end{cases}
\tag{6.3}
$$

Thus, equating the (6.3) to 0, and substituting $\delta_1(t) = 1 - \alpha(t)$ and $\delta_2(t) = -\beta(t)$:

$$
\begin{cases}
N^1 = e^{-(\alpha(t)+\beta(t)y^1(t))} \\
N^0 = e^{-(\alpha(t)+\beta(t)y^0(t))} \\
N^1 + N^0 = N \\
N^1 y^1 + N^0 y^0 = Y(t)
\end{cases}
$$

Substituting the first two equations in the third and rearranging, it becomes:

$$
e^{-\alpha(t)} = \frac{N}{e^{-\beta(t)y^1(t)} + e^{-\beta(t)y^0(t)}}
$$

which, substituted in the last of the (6.3), generates:

$$
e^{-\beta(t)y^1(t)}y^1(t) + e^{-\beta(t)y^0(t)}y^0(t) = Y(t)\frac{e^{-\beta(t)y^1(t)} + e^{-\beta(t)y^0(t)}}{N}
$$

Setting $\bar{y}(t) = Y(t)/N$, we obtain:

$$
(y^1(t) - \bar{y}(t))e^{-\beta(t)y^1(t)} + (y^0(t) - \bar{y}(t))e^{-\beta(t)y^0(t)} = 0
\tag{6.4}
$$

that is the equation of the *quantum anomalies*. Its l.h.s. measures the distance of the actual production from an ideal situation where $\mu = 0$ for all firms. $\beta(t)$ then emerges as an important synthetic indicator for the reduction in actual production due to the financial distress of the economic system. Its estimation as a solution for the equation (6.4) permits to obtain the theoretical probability of an agent to enter in state x or y, conditioned on the present value of N^1 and N^0. Since:

$$
N^j = e^{-\alpha(t)}e^{-\beta(t)y^j(t)}
$$

for $j = 0, 1$, then

$$
p^j = \frac{\hat{N}^j}{N} = \frac{e^{\beta(t)y^j(t)}}{Z}
\tag{6.5}
$$

---

[3] As demonstrated in Landini (2005, page 146) the first order conditions are also sufficient.

for $j = 0, 1$, where Z represents the partition function:

$$Z = e^{-y^1(t)\beta(t)} + e^{-y^0(t)\beta(t)}$$

The solution of equation (6.4) is the estimation for $\beta(t)$:

$$\beta(t) = \ln\left(-\frac{y^1(t) - \bar{y}(t)}{y^0(t) - \bar{y}(t)}\right)\left(y^1(t) - y^0(t)\right)^{-1} \qquad (6.6)$$

Taking equations (5.20) we obtain the following equivalences:

$$y^1(t) - \bar{y}(t) = -N^0\frac{2c\mu(t)}{r(r + 2c\mu(t))}$$

$$y^0(t) - \bar{y}(t) = N^1\frac{2c\mu(t)}{r(r + 2c\mu(t))} \qquad (6.7)$$

$$y^1(t) - y^0(t) = -\frac{2c\mu(t)}{r + 2c\mu(t)}$$

Substituting the (6.7) in the (6.6), β can be expressed as:

$$\beta(t) = \ln\left(\frac{N - N^1}{N^1}\right)\left(\frac{r + 2c\mu(t)}{2c\mu(t)}\right) \qquad (6.8)$$

As evidenced in Aoki (1996) and in Landini (2005, cap. 8), β, being dependent only on production levels, can be regarded as an economic index[4]. One can verify that, if the proportion of firms in state x increases, the variable's value drops, until it becomes less than 0, revealing an underperformance of the economy due to the risk of bankruptcy of heavily indebted firms. Indeed, the sign of β is then determined by the relative proportion of firms in the two states:

$$N^1 > N^0 \;\Rightarrow\; \beta < 0$$
$$N^1 < N^0 \;\Rightarrow\; \beta > 0$$
$$N^1 = N^0 \;\Rightarrow\; \beta = 0$$

The distance from 0 is basically due to the disproportion among the two occupation numbers (enforced by the interest rate effect), given that:

$$(N^1 \to N) \;\Rightarrow\; \beta \to -\infty$$
$$(N^1 \to 0) \;\Rightarrow\; \beta \to \infty$$

---

[4]On this point see also equations (3.20) in section 3.4 that permit to interpret this variable as the elasticity of the argument of the maximization to the production.

In this view β may be considered also as an index of the uncertainty of the system, given that, in the case of an approximately equal proportion of firms in the two states, the variable tends to 0.

Now equations (6.5) for $j = 0, 1$ can be expressed as:

$$p^0 = Z^{-1} \left(\frac{N - N^1}{N^1} \right)^{\frac{1}{2c\mu(t)}} \tag{6.9}$$

$$p^1 = Z^{-1} \left(\frac{N - N^1}{N^1} \right)^{\frac{r + 2c\mu(t)}{r2c\mu(t)}} \tag{6.10}$$

Therefore, the above equations represent the maximum likelihood estimations for the probability density function for a firm to enter in state x or y, where Z is the partition function.

6.2 Dynamics

6.2.1 Master equation

In the last section the statical probability for a firm of being in one of the two states has been quantified. The problem now is to specify the dynamics of the joint probabilities, in order to describe the evolution of the micro-states and, in this way, the stochastic evolution of the system. It is worth stressing that, in our system, we can observe two micro-states (x and y) and N macro states, since the state variable N_k can assume all the values in the interval $[0; N]$. Being interested in the macro dynamics, the analysis is focused on macro states and on the dynamics of their probability. I make use of the master (or Chapmann-Kolmogorov) equation, introduced in section 3.2 that permits to quantify the variation of probability fluxes in a small interval of time. The probability distribution of having N_k firms in state x in a given instant will follow this scheme:

$\frac{dP(N_k,t)}{dt}$ =(inflows of probability fluxes into x)-(outflows of probability fluxes out of x).

Then, it can be specified as follows:

$$\frac{dP(N_k,t)}{dt} = b(N_k - 1)P(N_k - 1) + d(N_k + 1)P(N_k + 1) + \\ - \{[(b(N_k) + d(N_k))P(N_k)]\} \tag{6.11}$$

with boundary conditions:

$$\begin{cases} P(N, t) = b(N_k)P(N_k - 1, t) + d(N)P(N, t) \\ P(0, t) = b(1)P(1, t) + d(0)P(0, t) \end{cases} \tag{6.12}$$

The conditions (6.12) ensure that the distribution functions consider only consistent values, that is to say $N_k \in [0; N]$.

6.2.2   Stationary points

As anticipated in section 3.2, in order to obtain an analytical solution for the master equation, it must be split into two components: the *drift*, that defines the tendency value of the mean of the state variable (N_k in the present case), and the *spread*, that expresses the probability of variations at aggregate level around the trend. These two components, in turn, permit to identify trend and fluctuations in production, quantifying the eventual equilibrium level of output and cycles around this trend. Then, one of the solution methods proposed by Aoki (2002), is applied. Firstly, by means of lead and lag operators, probability fluxes in and out the states can be treated as homogeneous. Then, the Taylor series expansion of the modified master equation isolates the drift equation and the Fokker-Plank equation. This method is maybe less sophisticated or less elegant compared to the above cited Van Kempen's and Kubo's methods but, at the same time, it appears to be, at least in the author's opinion, equally effective and hopefully clearer for what regards development and computational procedures.

In what follows, the state variable N_k of the master equation (6.11) is modified, formulating a hypothesis about its drift and spread components. First, attention is focused on the former, in order to isolate an eventual equilibrium dynamics. Then, by means of a further transformation of transition rates, an explicit solution for the equation of the spread is presented.

Drift and convergence dynamics

In this first approximation, attention is focused on the conditional mean of the fraction of agents that are in one of the two states (namely, state x, that is to say the more indebted firms), i. e. on the drift, in order to find an analytical formulation for this first component. By assumption, the number of firms in state x at a given moment is determined by two components: its expected mean (the drift), indicated by m; an additive fluctuations component around this value (the spread), indicated by s, that according to Aoki (2002), is set of order $N^{1/2}$. Therefore a new state variable is introduced and specified as follows:

$$N_k(t) = Nm + \sqrt{N}s \tag{6.13}$$

Solving the master equation (6.11), with respect to the trend component (see Appendix A) an explicit formulation for the macroscopic equation can be obtained:

$$\frac{dN_k}{dt} = -(\lambda + \gamma)N_k + \lambda N \tag{6.14}$$

This equation describes the dynamics of the drift. It can be interpreted as the long-run trend of the occupation number and, then, keeping all the other relevant variables unchanged, of the production. Now we can determine the stationary value of $N_k/N = n_k = m$ and the consequent steady state equilibrium of the economy, simply setting the r.h.s. of the (6.14) to 0:

$$m^* = \frac{\lambda}{\lambda + \gamma} = n_k^* \tag{6.15}$$

Once demonstrated the existence of an equilibrium point, a deeper insight on system's dynamics is needed in order to verify its convergence toward the equilibrium distribution. The solution of the differential equation (6.14) is:

$$m(t) = m(0)e^{-(\lambda+\gamma)t} + C \tag{6.16}$$

that, setting an opportune initial point $m(0)$, yields to:

$$m(t) = m^* + (m(0) - m^*)e^{-(\lambda+\gamma)t} \tag{6.17}$$

that verifies the convergence and the stability of the equilibrium since the second term goes to 0 as $t \to \infty$. This mechanism origins a smooth dynamics for the trend of production (entirely determined by the stochastic structure of firms' behaviors) around which fluctuations in the value of production, due to price shocks, take place. In this construction therefore the long-run path of the system is identified by its stochastic equilibrium, while business cycles are wholly due to idiosyncratic shocks on the demand side. We need then to model the spreading component.

Derivation of Fokker-Planck equation. Endogenous fluctuations

In this section a procedure for determining an explicit functional form for the spreading component is presented. This factor is endogenously generated by the system and its probability is quantified by means of *Fokker-Plank* equation. In this section, I apply the already introduced solution method for the master equation and obtain a treatable functional form for both drift and spreading component. Our economy can be fully described: we already know its macroscopic equation (that defines the trend) and now, solving the master equation (6.11) with respect to the spreading component, we can calculate the distribution of fluctuations. Details of calculation are reported in Appendix B.

The asymptotically approximated solution of the master equation is given by the following system of coupled equations:

$$\frac{dm}{d\tau} = \lambda m - (\lambda + \gamma)m^2 \qquad (6.18)$$

$$\frac{\partial Q}{\partial \tau} = [2(\lambda + \gamma)m - \lambda]\frac{\partial}{\partial s}(sQ(s)) + \frac{[\lambda m(1-m) + \gamma m^2]}{2}\left(\frac{\partial}{\partial s}\right)^2 Q(s) \quad (6.19)$$

Equation (6.18) is a deterministic ordinary differential equation called macroscopic equation which displays a logistic dynamics for the drifting component. Equation (6.19) is a second order stochastic partial differential equation called *Fokker-Planck equation* that drives the spreading component of the probability flow. Stationary point for equation (6.18) is the same as that already found for the equivalent equation (6.14) and presented in equation (6.15):

$$m^* = \frac{\lambda}{\lambda + \gamma}$$

Solution of the macroscopic equation can be obtained by direct integration of the (6.18), setting up a suitable Cauchy problem. Therefore:

$$m(\tau) = \frac{\lambda}{(\lambda + \gamma) - ke^{-\psi\tau}} \quad : \quad \begin{cases} k = 1 - \frac{m^*}{m(0)} \\ \psi = \frac{(\lambda+\gamma)^2}{\lambda} \end{cases} \qquad (6.20)$$

that, for the stationary condition $\dot{m} = 0$, returns again the (6.15).

Now a solution for the Fokker-Planck equation is identifiable. The aim is to find a solution for the probability flow in terms of $Q(s)$. The thus obtained solution will identify the stationary equilibrium since it is obtained by means of an asymptotic-approximation technique. Then, indicating with $\theta(s)$ the stationary probability for $Q(s)$ and setting the equilibrium condition $\dot{Q} = 0$ (that implies $\dot{\theta} = 0$), it is possible to obtain:

$$-[2(\lambda + \gamma)m^* - \lambda]\, s\theta(s) = \frac{[\lambda m^*(1 - m^*) + \gamma m^{*2}]}{2}\left(\frac{\partial}{\partial s}\right)\theta(s) \qquad (6.21)$$

Rewriting (6.21) more conveniently as:

$$\frac{2[\lambda - 2(\lambda + \gamma)m^*]}{[\lambda m^*(1 - m^*) + \gamma m^{*2}]}s = \frac{1}{\theta(s)}\left(\frac{\partial}{\partial s}\right)\theta(s) \qquad (6.22)$$

and integrating it with respect to s, we obtain:

$$\log\theta(s) = C + \frac{\lambda - 2(\lambda+\gamma)m^*}{\lambda m^*(1-m^*) + \gamma m^{*2}}s^2$$
$$\Updownarrow \qquad\qquad\qquad (6.23)$$
$$\theta(s) = C\exp\left(\frac{\lambda - 2(\lambda+\gamma)m^*}{\lambda m^* + (\gamma-\lambda)m^{*2}}s^2\right)$$

where C is the constant of integration. Then, substituting $m^* = \lambda/(\lambda + \gamma)$, we get the final result:

$$\theta(s) = C \exp\left(-\frac{s^2}{2\sigma^2}\right) \quad : \quad \sigma^2 = \frac{\lambda\gamma}{(\lambda + \gamma)^2} \qquad (6.24)$$

which looks like a Gaussian density whose parameters are dependent uniquely on the transition rates λ and γ. Therefore, by means of this last method, the dynamics of both drift and spread can be modeled as endogenous.

At this point it is possible to deepen the analysis, stating further hypothesis about the the spread s in order to better define it. We then model this factor referring it to the existing empirical evidence about micro-fluctuations and their effect on macro-volatility.

Using Gabaix's results on aggregate volatility

In a recent paper, Gabaix shows that volatility of aggregate production is proportional on GDP's size (Gabaix, 2005). At the root of his work is the evidence, that dates back to Mandelbrot (1963) and Ijiri and Simon (1977), that production units have a fat tail distribution and, precisely, as more recently detected by a number of interdisciplinary studies[5], a Levy-stable distribution with power law tails. This stylized fact has non negligible consequences when one analyzes the relationships between industrial structures and GDP's time series of an economic system[6]. Some of the effects of firms' distribution's features and dynamics on business cycles have been previously analyzed, among others, by Fujiwara et al. (2004) and Gaffeo et al. (2003).

But the most important implication of fat tailed distribution, for the aim of this work, is that, following Gabaix, idiosyncratic firm level shocks lead to non-trivial aggregate fluctuations. This is very much in line with the approach here proposed and permits to add new elements on our theoretical construction. The model proposed by Gabaix demonstrates that, under specific hypothesis on the underlying firms' distribution (that are consistent with empirical results), the micro volatility (for example, variations in firms' production levels) and the macro volatility (i. e. business cycles) have the same power law distribution. Formally:

$$\sigma^{firms}(S) \sim \sigma^{GDP}(S) \sim S^{-\alpha}$$

---

[5]See, as a matter of example, Axtell (2001) and Okuyama et al. (1999).

[6]Note that the nth moment of a Levy stochastic process exists if and only if the tail parameter of its distribution is equal or bigger than n. The tail parameter usually found in empirical research is around 1. This should be enough to cast some doubts on the possibility to use representative agent hypothesis in business cycles (or generally in macroeconomic) analysis.

where σ represents the volatility and S is the size, while the power law shape parameter α has a value included in the interval $(0; 1/2]$. These theoretical results have been validated by the empirical analysis reported in the same paper.

Thus, in order to set an explicit formulation for s (and then quantify the probability of output fluctuations in the model), equation (6.13) is reformulated in the following way:

$$N_k = Nm + \sqrt{N}\tilde{s} \qquad (6.25)$$

with:

$$\tilde{s} = f(s) = s^{-1/\alpha}S \qquad (6.26)$$

Now the new information can be embodied in the probability function for fluctuations. Considering that in equation (6.23):

$$C = \left[\int^s \exp\left(\frac{\lambda - 2(\lambda + \gamma)m^*}{\lambda m^* + (\gamma - \lambda)m^{*2}} s^2 \right) ds \right]^{-1}$$

and taking the opportune partition function Z, we obtain:

$$p(\tilde{s}) \propto \frac{2\alpha}{Z}\tilde{s}^{3\alpha}exp\left(-\frac{\tilde{s}^{2\alpha}}{Z} \right) \qquad (6.27)$$

that is a Weibull-type distribution function. It appears consistent with the evidence that we found, in a previous paper, with regard to a sample of 16 industrialized countries over a 120 years time span (Di Guilmi et al., 2004). Analyzing business fluctuations' distribution for all the countries included in the sample, individuated as the cumulative percentage deviations from the trend (defined by means of Hodrick-Prescott filtration of the raw series) within two turning points, we found that the Weibull distribution returns the best fit.

In the same way, it is possible to get a measure of cumulative aggregate fluctuations that may be more appropriate to the nature of the generative mechanism here introduced and that allows to take into explicit consideration the duration, in terms of units of time, of the phases. We can then make a comparison with a different measure that has been defined by Di Guilmi et al. (2005) as *steepness*, that is the ratio among the absolute value of the cumulative percentage points of peak-to-trough and trough-to-peak output gap for recessions and expansions, respectively, to the time duration of the phase (expressed in number of periods). Therefore, we have to introduce a different hypothesis for the spread in order to identify a variable suitable for comparison. In order to get the most general possible measure I approximate the duration of the phases with their distribution, as done above for the spread s. There are some recent (and partially contradictory) results as regards distribution of phases' duration. For Ormerod and Mounfield (2001) durations of contractions follow a power law distribution while, according to Wright (2003), data are better fitted by an exponential law. For Ausloos et al. (2004) the duration of recession phases is power law distributed, while expansions duration distribution is exponential. I performed an analysis on the USA phase durations using a public NBER data set that covers monthly data from December 1854 to November 2001[7]. It returns an acceptable fit for power law distribution for expansions and contractions, above a given threshold of duration, taking into account that "a cycle is designated in the NBER methodology only if it has achieved a certain maturity" (Diebold and Rudebusch, 1990). Therefore let us define the duration distribution as:

$$p(D) = bD^{-a} \tag{6.28}$$

Since the model works in continuous time, we can write:

$$s\frac{dP}{dt} = sS^{-\alpha}$$

The integration with respect to time returns:

$$sP_s(t) = sS^{-\alpha}t + C \tag{6.29}$$

Now, setting $b = 1$ and substituting to t its theoretical value DD^{-a}, the (6.29) becomes:

$$sP_s(t) = sS^{-\alpha}D^{1-a} \tag{6.30}$$

It is possible to define the spread $\tilde{s}$ as:

$$\tilde{s} = f(s, S, D) = sSD^{\frac{a-1}{\alpha}} \tag{6.31}$$

Proceeding as above it is possible to obtain a distribution analogue to the (6.27) and comparable to one identified by Di Guilmi et al. (2005).

---

[7]Results are detailed in section 10.1.

Appendix A: Macroscopic equation

Master equation (6.11) has now to be modified in order to consider the new variable (6.13). It can be expressed as $\dot{Q}(s)$, a function of s, and then it becomes:

$$\dot{P}(N_k) = \frac{\partial Q}{\partial t} + \frac{ds}{dt}\frac{\partial Q}{\partial s} = \dot{Q}(s) \tag{6.32}$$

with transition rates reformulated in the following way:

$$b(s) = \lambda \left[N - Nm - \sqrt{N}s \right] \tag{6.33}$$

$$d(s) = \gamma \left[Nm + \sqrt{N}s \right] \tag{6.34}$$

Since:

$$\frac{ds}{dt} = -N^{1/2}\frac{dm}{dt} \tag{6.35}$$

equation (6.32) can be expressed as:

$$\dot{Q}(s) = \frac{\partial Q}{\partial t} - N^{1/2}\frac{\partial Q}{\partial s}\dot{m} \tag{6.36}$$

Now I rewrite again the master equation (6.11) and the transition rates by means of lead and lag operators. The use of these operators allows to express the master equation in a more treatable form, making the two probability flows (in and out) homogeneous. Specifically the transition probabilities (5.3) become:

$$H[d(N_k)P(N_k, t)] = d(N_{k+1})P(N_{k+1}) \tag{6.37}$$
$$H^{-1}[b(N_k)P(N_k, t)] = d(N_{k-1})P(N_{k-1}) \tag{6.38}$$

so that the master equation will be expressed in this way:

$$\dot{Q}(s) = (H - 1)[d(s)Q(s)] + (H^{-1} - 1)[d(s)Q(s)] \tag{6.39}$$

Using the modified transition rates (6.37) and expanding the so obtained master equation in inverse powers of s to the second order we get:

$$
\begin{aligned}
&N^{-1}\frac{\partial Q}{\partial \tau} - N^{-1/2}\frac{dm}{d\tau}\frac{\partial Q}{\partial s} = \\
&N^{-1/2}\left(\frac{\partial}{\partial s}\right)[d(s)Q(s)] + N^{-1}\frac{1}{2}\left(\frac{\partial}{\partial s}\right)^2[d(s)Q(s)] + \\
&-N^{-1/2}\left(\frac{\partial}{\partial s}\right)[b(s)Q(s)] + N^{-1}\frac{1}{2}\left(\frac{\partial}{\partial s}\right)^2[b(s)Q(s)] + \dots \\
&= N^{-1/2}\left(\frac{\partial}{\partial s}\right)[(d(s) - b(s))Q(s)] + N^{-1}\frac{1}{2}\left(\frac{\partial}{\partial s}\right)^2[(b(s) + d(s))Q(s)] + \dots
\end{aligned}
\tag{6.40}
$$

At this point, in order to match components of the same orders of powers of N between and equations (6.32) and (6.40), we need to rescale the variable $\tau = tN$. Knowing that:

$$d(s) - b(s) = (\lambda + \gamma)(Nm + \sqrt{N}s) - \lambda N = (\lambda + \gamma)N_k - \lambda N$$
$$d(s) + b(s) = (\lambda - \gamma)(Nm + \sqrt{N}s) + \lambda N = (\lambda - \gamma)N_k + \lambda N$$

and taking the first order derivatives, up to the second order, it is possible to obtain what Aoki (2002) defines as *diffusion approximation*:

$$N^{-1}\frac{\partial Q}{\partial \tau} - N^{-1/2}\frac{dm}{d\tau}\frac{\partial Q}{\partial s} =$$
$$(\lambda + \gamma)Q(s) + N^{-1/2}(d(s) - b(s))\left(\frac{\partial}{\partial s}\right)Q(s) + N^{-1}\frac{1}{2}(b(s) + d(s))\left(\frac{\partial}{\partial s}\right)Q(s) \tag{6.41}$$

Equating the terms of order $N^{-1/2}$, for the polynomial identity principle, we get:

$$N^{-1/2}\frac{dm}{dt}\frac{\partial Q}{\partial s} = -N^{-1/2}(b(s) - d(s))\left(\frac{\partial}{\partial s}\right)Q(s)$$

Finally, an explicit formulation for the macroscopic equation is identified:

$$\frac{dN_k}{dt} = -(\lambda + \gamma)N_k + \lambda N$$

Appendix B: Fokker-Planck equation

Applying the polynomial identity principle to equation (6.41) for powers of N of order $-1/2$, we found the deterministic differential equation (6.14) that represents our macroscopic component. Repeating the procedure for powers of N of order -1 we finally get a formulation for the *Fokker-Planck* equation:

$$N^{-1}\frac{\partial Q}{\partial t} = -b'(m)s\frac{\partial Q}{\partial s} + b(m)\frac{1}{2}\left(\frac{\partial}{\partial s}\right)^2 Q(s) - b'(m)sQ(s)$$
$$+d'(m)s\left(\frac{\partial}{\partial s}\right)Q(s) + d(m)\frac{1}{2}\left(\frac{\partial}{\partial s}\right)^2 Q(s) - d'(m)Q(s) \tag{6.42}$$
$$= (d'(m) - b'(m))\left(\frac{\partial}{\partial s}\right)Q(s) + \frac{1}{2}(d(m) + b(m))\left(\frac{\partial}{\partial s}\right)^2 Q(s)$$

As before, we want to find an explicit form for the generic asymptotic approximated solution to the master equation, given by the solution of the following coupled dynamical system of equations:

$$\begin{cases} \dfrac{dm}{d\tau} = \rho(m) \\ \dfrac{\partial Q}{\partial \tau} = -\rho'(m)\left(\dfrac{\partial}{\partial s}\right)(sQ(s)) + \dfrac{1}{2}\alpha(m)\left(\dfrac{\partial}{\partial s}\right)^2 Q(s) \\ \text{s.t.} \quad \rho(m) = b(m) - d(m), \quad \alpha(m) = b(m) + d(m) \end{cases} \tag{6.43}$$

This is a generic (and implicit) solution since no functional form for transition rates has been specified until now. In order to arrive at an explicit solution I introduce a modification in the transition rates of equations (5.3), supposing that the probability η is equal to the observed frequency of firms occupying state x. The new transition rates are then:

$$
\begin{cases}
b_n = r(N_k + 1|N) = \zeta \dfrac{N_k}{N} \dfrac{N - N_k}{N} \\
d_n = l(N_k - 1|N) = \iota \dfrac{N_k}{N} \dfrac{N_k - 1}{N}
\end{cases}
\tag{6.44}
$$

where the factor $\frac{N_k}{N}$ allows to interpret, respectively, the probability transition ζ as a constant of proportionality between the birth rate per individual and the deviation from the upper bound $N - N_k$ and the probability transition ι as a constant of proportionality between the death rate per individual and the deviation from the lower bound or $N_k - 1$. Given that, one can set the two functions:

$$
\lambda(N_k) = \lambda \frac{N_k}{N}
\tag{6.45}
$$

$$
\gamma(N_k) = \gamma \frac{N_k}{N}
\tag{6.46}
$$

for birth and death rate depending on the state specific volume of agents in a given state. Using the above specified transition rates it is then possible to solve the master equation (6.11), in order to render explicit both the macroscopic and Fokker-Planck associated equations. Substituting the transition rates (6.45) in the master equation (6.11) and collecting terms with λ and γ, after some simple but tedious algebraic passages, we obtain:

$$
\frac{dP}{dt} = N^{-2} \left\{ \gamma \left[N_k(N_k + 1)L(P) + 2nP \right] + \right.
$$
$$
\left. -\lambda \left[(N_k - 1)(N - N_k + 1)L^{-1}(P) + (N - 2n + 1)P \right] \right\}
\tag{6.47}
$$

where $L(P)$ and $L^{-1}(P)$ are lead and lag operators, reformulated in the following way according to Aoki (1996) and Landini and Uberti (2008):

$$
L(P) = \sum_{z=1}^{\infty} \frac{N^{-z/2}}{z!} \left(\frac{\partial}{\partial s} \right)^z Q(s)
\tag{6.48}
$$

$$
L^{-1}(P) = \sum_{z=1}^{\infty} \frac{(-)^z N^{-z/2}}{z!} \left(\frac{\partial}{\partial s} \right)^z Q(s)
\tag{6.49}
$$

Using the above indicated operators into equation (6.47), it becomes:

$$
\frac{dP}{dt} = N^{-2} \left\{ \sum_{z=1}^{\infty} [D(N_k) + (-)^z B(N_k)] \frac{N^{-z/2}}{z!} \left(\frac{\partial}{\partial s} \right)^z Q(s) \right\} +
$$
$$
+ N^{-2} \left\{ [2\gamma N_k - \lambda(N - 2n + 1)] Q(s) \right\}
\tag{6.50}
$$

where:

$$\left\{ \begin{array}{l} B(N_k) = \lambda(N_k - 1)(N - N_k + 1) = \lambda N(N_k - 1) - \lambda(N_k - 1)^2 = B(m) \\ D(N_k) = \gamma N_k(N_k + 1) = D(m) \end{array} \right.$$

$$(6.51)$$

The specification of the drift displayed in equation (6.13) implies that:

$$\left\{ \begin{array}{l} N_{k+1} = Nm + \sqrt{N}(s + N^{-1/2}) \\ N_k = Nm + \sqrt{N}s \\ N_{k-1} = Nm + \sqrt{N}(s - N^{-1/2}) \end{array} \right.$$

$$(6.52)$$

Using these specifications in equation (6.51), it turns out to be:

$$\left\{ \begin{array}{l} B(m) = \lambda \left[N^2 m(1 - m) + N^{3/2} s(1 - 2m) + N(2m - s^2 - 1) + 2N^{1/2} s - 1 \right] \\ D(m) = \gamma \left[N^2 m^2 + N^{3/2} 2ms + N(m + s^2) + N^{1/2} s \right] \end{array} \right.$$

$$(6.53)$$

Now, we expand to the second order equation approximation equation (6.50), that is to say for $z = 1, 2$, getting:

$$\left\{ \begin{array}{ll} z = 1 : & N^{-1/2}\left[D(m) - B(m)\right] = \\ & N^{3/2}\left[-\lambda m(1 - m) + \gamma m^2\right] + N\left[2ms(\lambda + \gamma) - \lambda\langle s\rangle\langle s\rangle\right] + \\ & +N^{1/2}\left[s^2(\lambda + \gamma) + (\gamma - 2\lambda)m + 1\lambda\right] - 2\lambda s + N^{-1/2}(\gamma s - \lambda) \\ z = 2 : & \frac{N-1}{2}\left[D(m) + B(m)\right] = \\ & N\left[\lambda m(1 - m) + \gamma m^2\right] + N^{1/2}\left[2ms(\lambda + \gamma) - \lambda s\right] + \\ & +\left[s^2(\gamma - \lambda) + m(\gamma + 2\lambda) - \lambda\right] + N^{-1/2}2\lambda s + N^{-1}(\gamma s + \lambda) \end{array} \right.$$

$$(6.54)$$

and substituting it into equation (6.50), we arrive at the following approximated master equation:

$$\begin{aligned} \frac{dP}{dt} = \ & \left\{ N^{-1/2}\left[-\lambda m(1 - m) + \gamma m^2\right] + N^{-1}\left[2ms(\lambda + \gamma) - \lambda s\right] \right\} \frac{\partial}{\partial s}Q(s) + \\ & + \left\{ -N^{-3/2}\left[s^2(\lambda + \gamma) + (\gamma - 2\lambda)m + \lambda\right] - N^{-2}2\lambda s + \right. \\ & \left. -N^{-5/2}(\gamma s - \lambda) \right\} \frac{\partial}{\partial s}Q(s) + \frac{1}{2}\left\{ N^{-1}\left[\lambda m(1 - m) + \gamma m^2\right] + \right. \\ & \left. +N^{-3/2}\left[2ms(\lambda + \gamma) - \lambda s\right] \right\} \left(\frac{\partial}{\partial s}\right)^2 Q(s) + \\ & +\frac{1}{2}\left\{ N^{-2}\left[s^2(\gamma - \lambda) + m(\gamma + 2\lambda) - \lambda\right] + \right. \\ & \left. +N^{-5/2}\lambda s + N^{-3}(\gamma s + \lambda) \right\} \left(\frac{\partial}{\partial s}\right)^2 Q(s) + \\ & + \left\{ N^{-1}\left[2m(\lambda + \gamma) - \lambda\right] + N^{-3/2}\left[2s(\lambda + \gamma)\right] - N^{-2}\lambda \right\} \end{aligned}$$

$$(6.55)$$

Considering that:

$$\frac{dP}{dt} = \frac{\partial Q}{\partial t} - N^{-1/2}\frac{dm}{dt}\frac{\partial Q}{\partial s}$$

$$(6.56)$$

in order to match the higher order terms in powers of N, we have to rescale time as $t = N\tau$:

$$\frac{dP}{dt} = \frac{\partial Q}{\partial t} - N^{-1/2}\frac{dm}{dt}\frac{\partial Q}{\partial s} \Leftrightarrow N^{-1}\frac{dP}{d\tau} = N^{-1}\frac{\partial Q}{\partial \tau} - N^{-1/2}\frac{dm}{d\tau}\frac{\partial Q}{\partial s} \quad (6.57)$$

Then we have to equal the thus obtained two formulations for the master equation: equation (6.55) and equation (6.57). This can be done matching the terms that have the same power of N. Then we collect terms of order N^{-1} in equation (6.55) so as to match them with $\partial Q/\partial \tau$ of equation (6.57) and those of order $N^{-1/2}$ to set them equal to $N^{-1/2}\dot{m}\frac{\partial Q}{\partial s}$. All the other terms asymptotically vanish as $N \to \infty$. In this way we get:

$$-N^{-1/2}\frac{dm}{d\tau}\frac{\partial Q}{\partial s} = -N^{-1/2}\left[\lambda m(1-m) - \gamma m^2\right]\frac{\partial}{\partial s}Q(s) \quad (6.58)$$

$$\begin{aligned} N^{-1}\frac{\partial Q}{\partial \tau} = \ & N^{-1}[2m(\lambda + \gamma) - \lambda]\frac{\partial}{\partial s}(sQ(s)) + \\ & + \frac{N^{-1}}{2}[\lambda m(1-m) + \gamma m^2]\left(\frac{\partial}{\partial s}\right)^2 Q(s) \end{aligned} \quad (6.59)$$

Finally, the system of coupled equations (6.43) can be expressed in an explicit form, that represents the asymptotically approximated solution of the master equation:

$$\frac{dm}{d\tau} = \lambda m - (\lambda + \gamma)m^2$$

$$\frac{\partial Q}{\partial \tau} = [2(\lambda + \gamma)m - \lambda]\frac{\partial}{\partial s}(sQ(s)) + \frac{[\lambda m(1-m) + \gamma m^2]}{2}\left(\frac{\partial}{\partial s}\right)^2 Q(s)$$

Chapter 7

Stationary distributions

This chapter closes the model by applying the results obtained in chapter 6 to the analytical structure of our economy, in order to get an aggregate supply curve (section 7.1), its stochastic dynamics (section 7.2) and a new formulation of transition rates that embodies all the stochastic determinants of the potential evolution of the economy.

7.1 Aggregation and origin of business cycles

The aggregate output of the system is given by:

$$Y(t) = \frac{N^1}{r + 2c\mu(t)} + \frac{N^0}{r} = N \left[y^0(t) - n^1 \left(y^0(t) - y^1(t) \right) \right] \qquad (7.1)$$

It is straightforward to note that the condition for the steady state of production is the same condition for the stationary state (in stochastic terms) of the system. Given the difference in firms' optimal production, temporal fluctuations in the level of $Y(t)$ are due to changes in the levels of N^1 and N^0. Precisely, we will have a recession if the value of N^1 increases and an upturn if it decreases.

Having already quantified the equilibrium distribution of the drift, it is possible now to obtain the steady state value of aggregate production, simply by substituting equation (6.15) in the (7.1):

$$Y^e = N \left[y^0(t) - \frac{\lambda}{\lambda + \gamma} \left(y^0(t) - y^1(t) \right) \right] \qquad (7.2)$$

Since the aggregate production function depends on m, also its dynamic will be convergent to a stationary level. Fluctuations then come out to be dependent on the transition rates λ and γ and on the differences in firms level of production. The dynamics of these two factors are studied in the next sections.

7.2   Equilibrium distribution and critical points

Equating master equation to 0, it is possible to obtain the Kolmogorov condition that equates the probability fluxes entering a state with the fluxes coming out from that state. Using equations (5.3) and (6.14) the master equation in intensive form becomes:

$$\dot{m} = \zeta\eta - \{\iota[1 - \eta] + \zeta\eta\} m \qquad (7.3)$$

Setting $\dot{m}$ to 0 and rearranging, the stationary configuration of the system can be expressed as:

$$\dot{m} = 0 \Rightarrow \frac{\eta}{1 - \eta} = \frac{\iota m}{\zeta(1 - m)} \qquad (7.4)$$

As illustrated in section 3.1, in a Markovian space, we can make use of Brook's lemma (Brook, 1964) that defines local characteristic of this kind of chain. It takes into account all the possible states (i.e. the possible number of occupancies of each state, from 0 to N) of the system and states that, for detailed balance condition to hold for each couple of states, we have:

$$P^e(N_k) = P^e(N(0)) \left(\frac{\iota}{\zeta}\right)^{N_k} \binom{N}{N_k} \prod_{h=1}^{k} \frac{\eta(N - N_h)}{(1 - \eta(N_h))} \qquad (7.5)$$

And then, thanks to the Hammersley and Clifford theorem, the expected stationary probability of the markovian process for N^1, when detailed balance holds, can be expressed by:

$$P^e(N_k) \propto Z^{-1} e^{-\beta N U(N_k)} \qquad (7.6)$$

where $U(N_k)$ is the Gibbs potential and Z is the partition function. β is the same parameter calculated in section 3.4 and therefore, may be interpreted as an inverse measure of the uncertainty of the system. The above formulation allows us to express explicitly the values of the a-priori probability in Boltzmann-Gibbs form (Aoki, 2002; Aoki and Yoshikawa, 2006):

$$\boxed{\begin{aligned} \eta(N^1) &= N^{-1} e^{\beta g(N^1)} \\ 1 - \eta(N^1) &= N^{-1} e^{-\beta g(N^1)} \end{aligned}} \qquad (7.7)$$

so that:

$$e^{\beta g(N^1)} + e^{-\beta g(N^1)} = N \tag{7.8}$$

where $g(N^1)$ is a function that evaluates the relative difference in the outcome as a function of N^1. It is easy to verify that large values of β associated with positive values of $g(N^1)$ cause $\eta(N^1)$ to be larger than $1 - \eta(N^1)$, making transition from state y to state x more likely to occur than the opposite one.

To get a deeper insight of the interpretation of β let us formulate explicitly the potential equation, previously introduced in implicit form (equation (3.1)). In binary models, as the one treated here, and for great N, the equation of the potential is:

$$U(N^j) = -2 \int_0^{N^j} g(z)dz - \frac{1}{\beta} H(\underline{N})$$

where $H(\underline{N})$ is the Shannon entropy with $H(\underline{N}) : \underline{N} = (N^1, N^0)$. In order to identify the stationary points of probability dynamics we need to identify its peak (if it exists). β is an inverse multiplicative factor for entropy: this implies that, for very large values of it, the entropy component does not play any role. On the contrary, as β approaches 0, the weight of the entropy component grows. In terms of the present model, a relative high value of β means that uncertainty at macroeconomic level is low, with few possible configurations of the occupation numbers (Aoki, 1996, pp. 55 and followings). Therefore for values of β around 0, and a more relevant volatility in the system, in order to identify the peak of probability dynamics we need to find the local minimum of the potential. It is possible now to find the stationary probability for our system's dynamics.

Aoki (2002) shows that the points in which the potential is minimized are also the critical points of the aggregate dynamics of $P(N^j)$. Deriving the potential respect to N^1 and then setting $U' = 0$:

$$g(N^1) = -\frac{1}{2\beta} \frac{dH(\underline{N})}{dN^1} = -\frac{1}{2\beta} ln\left(\frac{N^1}{N - N^1}\right) \tag{7.9}$$

and using the formulation of β reported in equation (6.6), we get an explicit formulation for $g(N^1)$ in stationary condition:

$$g(N^1) = \frac{y^0 - y^1}{2} \tag{7.10}$$

that quantifies the mean difference (for states) of the outcome. From equation (7.9) it follows that the point of local minimal of the potential is given by:

$$U' = 0 \Rightarrow e^{2\beta g(N^1)} = \frac{N^1}{N - N^1} \tag{7.11}$$

that, divided by N, if the rates of entries and exits are equated, (i. e. if $\iota = \zeta$), reproduces exactly the (7.4). Therefore, making use of the (7.7) we can write:

$$P(N^1) = \frac{e^{\beta g(N^1)}}{e^{\beta g(N^1)} + e^{-\beta g(N^1)}} \tag{7.12}$$

that is the maximum likelihood estimation of Gibbs probability density of the number of firms in state x.

Let us analyze now, in greater detail, the different behavior of the stationary distribution for different values of β:

1. $\boxed{\beta \to \infty}$ In this case, given equation (7.9), the critical points in which the potential is minimized are also the zeros of the function $g(N^1)$. Indeed:

$$U'(m^*) = -2g(m^*) = 0 \tag{7.13}$$

   This confirms that β may be interpreted as an inverse index of uncertainty. Indeed, considering equation (7.10):

$$g(N^1) = 0 \Leftrightarrow y^0 - y^1 = 0$$

   Under the specified conditions, there is no uncertainty in the system, since any firm can go bankrupted and, therefore, the level of production is unambiguously determined. Indeed, taking equation (6.8), the value of β can go to infinity if $N^0 \to N$ or if $\mu \to 0$, since both situations imply a convergence among the different targets of production at micro level and, then, a minimum degree of uncertainty in the system;

2. $\boxed{\beta \to 0}$ In this second *scenario*, in order to identify the critical points of the dynamics, a further deepening is needed. Taking again equation (6.8), it is straightforward to note that β can go to 0 if and only if $\frac{N^1}{N-N^1} \to 1$, that is to say, if the system is populated, in the same proportion, of firms in state x and of firms in state y. But this is not informative about the local minimum of the potential since, given that as β approaches 0, $g(N^1)$ goes to infinity. Therefore, in order to determine critical points a hazard function analysis is performed. Cox and Miller (1996) define the hazard function for a stochastic variable x as:

$$h(x) = \lim_{v \to 0^+} \frac{Pr(x < X < x + v | x < X)}{v} = -\left(\frac{d}{dx}\right) \log(1 - F(x))$$

   Assuming that the function $Pr(x < X)$ has the same function of the first-passage time of a standard Brownian motion (Grimmett and

Stirzaker, 1992), we can specify a functional form for $F(x)$:

$$F(x) = \left[1 + e^{-2\beta x}\right]^{-1} \tag{7.14}$$

The hazard function (7.14) in terms of m becomes:

$$F(m) = \left[1 + e^{-2\beta m}\right]^{-1} \Rightarrow h(m) = \frac{2\beta}{1 + e^{-2\beta m}} \tag{7.15}$$

Now, we have to calculate the probability that a firm passes from one state to another in response to a small variation in the difference of relative production, conditional on the current difference between y^0 and y^1, quantified by $g(m)$ from equation (7.10). Considering that $F(m) = P(g(m^*) < m)$:

$$h(m) = \lim_{v \to 0^+} \frac{Pr(m < g(m^*) < m + v | m < g(m^*))}{v} =$$
$$= -\left(\tfrac{d}{dm}\right) log\left(1 - F(m)\right) \tag{7.16}$$

We can then rewrite the conditional hazard function in the following way:

$$P(v \leq y^1 - y^0 | m^*) = \left[1 + e^{-2\beta g(m^*)}\right]^{-1} \tag{7.17}$$

and then:

$$h(y^1 - y^0 | m^*) = \frac{2\beta \eta(m^*)}{1 + e^{-2\beta g(m^*)}} \tag{7.18}$$

In the simplest case, supposing $\eta(m^*) = m^*$, we finally obtain:

$$h(y^1 - y^0 | m^*) = 2\beta m^* \tag{7.19}$$

Therefore, for values of β close to 0, the critical point of probability dynamics is a value of m^* approximately equal to β itself. In other words, β may be considered as the conditional hazard rate in the range where β is small. The potential then is minimized for a fraction m^* of firms in state x when the value of the conditional hazard function is approximately equal to β. This is not a surprising result, taking into account the similarity of equation (7.19) with the (7.13).

7.3   Concluding remarks

7.3.1   A reinterpretation of transition rates

It is now possible, summing up all the results obtained up to this point, to get a more informative formulation for transition rates. Recalling equations (5.17), (5.18) and (7.7) and substituting them into equations (5.3), which I report here for the sake of clarity:

$$\lambda = \eta\zeta$$
$$\gamma = (1 - \eta)\iota$$

we obtain a formulation of the transition rates λ and γ that express them as a function of the stochastic parameters β and $g(N^1)$ and of the shock price thresholds:

$$\lambda = \left[N^{-1}e^{\beta g(N^1)}\right]\left(2\bar{u}^\zeta - 1.5\right) \tag{7.20}$$

$$\gamma = \left[N^{-1}e^{-\beta g(N^1)}\right]\left(-2.5\bar{u}^\iota + 1.5\right) \tag{7.21}$$

The transition rates, and, by means of them, the proportion of firms in each state and, finally, the production level, appear to be dependent on a structural component, represented by the first factor, which dynamics and convergence have been analyzed in the previous paragraph, and on a conjuncture factor, linked to the shocks in prices and to the relative financial soundness/weakness of firms within the two clusters. As in the original models of Greenwald and Stiglitz, keeping all the other parameters fixed, business fluctuations appear to be originated, on one hand, by small idiosyncratic shocks on demand which effects are amplified by financial fragility. On the other hand, consistently with the other inspiring work of Delli Gatti et al. (2005), there is a second order effect, due to the interaction of firms in the system, that can be defined as financial contagion. But, differently from that model, interaction here is not indirectly made effective by bank(s), but is modeled with mean-field equations that directly involve firms' demography dynamics. Anyway, in both works, the mechanism of interaction makes firms react to the degree of financial fragility of the system (there through interest rate, here through the proportion of firms risking bankruptcy).

This result shows how a construction that considers heterogeneity and interactions of firms permits to model the dynamics of the system, at least from the particular point of view adopted here, with a bit more realism than a construction based on the representative agent. Moreover it appears to be an effective instrument to model a system that permits to enhance the investigation on the links among financial weakness, recession and growth dynamics, along the lines of Minsky's and New Keynesian approaches. Finally, it has to be remarked that, analogously to Gallegati (2002), it is quite difficult and inappropriate to identify the long period trend as a natural level of production, since, even though the proportions of firms in the two states asymptotically tend to a stationary value, this dynamics is purely stochastic.

7.3.2 On the results of stochastic analysis

The model was closed in section 6.2 by identifying the two coupled equations that fully describe economic dynamics. The analysis performed in this chapter, and in particular in section 7.2, goes a step further, identifying the endogenous value of the probability η, a stationary distribution function for macro states and the critical points of the dynamics. The aim of this last part, besides providing an integrated representation of the stochastic system, is to expand on the meaning of the parameters used in deriving the stochastic dynamics (in particular parameter β) and to depict a more complete representation of mean-field interaction.

The study of section 7.2 stresses the relevance of β for identifying the stochastic evolution of macro states. As explained in Aoki and Yoshikawa (2006) and Landini (2005), this variable represents a synthetic and useful indicator for evaluating the state of the economy at each point in time, since it is informative not only about the quote of firms in each state but, in general, about stability of the macro state and about the probable direction of its evolution. Recalling that it has been introduced as a transformation of the Langrange multiplier in the MaxEnt problem (section 6.1), it can be regarded as an index of the entropic level of the system in terms of intensity of interchange among states or, put in different words, of the sensitivity of statistical dynamics to microeconomic evolution. As shown in greater detail in part III, estimations of β can be calculated and analyzed both in virtual and in real economies, permitting an analysis that actually links micro variables (the distribution of firms according to their financial situation in the present study) and macro evidence (aggregate production).

On the other hand, the set of analytical instruments built and used in previous sections permits a full representation of dynamic interaction. In particular, the joint consideration of the indicator β and of the function $g(N^j)$ highlights the effect of economic conjuncture on firms' stochastic evolution. Specifically, these factors measures the effect of financial fragility on stationary probability originated by two variables: the probability of failure μ, that reduces the targeted output for a class of firms; the percentage of firm in each state that influences the probability of change of state (i. e. to be safe or not from failure), defining in this way the condition of relative financial distress or soundness of the system. As Aoki and Yoshikawa (2006) point out:

> "It is the combination of economic behavior ($g(x)$ function) and uncertainty (small β) that generates multiple minima for the potential function, and accordingly multiple equilibria in the economy".

These effects determine the stationary equilibrium of the probability dynamics through the Gibbs potential, and, therefore, the final level of output that, in turn, influences the transition probabilities, making effective the feedback mechanisms that characterize our economic system as complex.

It should appear clear that this structure can be defined for any stochastic system, independently from the number of micro states. A modeling of full heterogeneity can be built on the same scheme designed in this work, which provides a first, simplified version in order to evidence the stochastic procedure of aggregation and its potentials in economic analysis.

Chapter 8

Modeling a monetary policy

In this chapter the model is enriched by the introduction of monetary policies. Two different approaches are suggested. In the first section, a monetary authority makes use of a Taylor rule to modify the statical probability of firms to be in one of the two states and, therefore, modify aggregate production. It is demonstrated that, using aggregate indicators to reach a pre-determined level of output, the monetary authority cannot modify the stochastic behavior of the system. In section 8.2, I use a different approach, showing that the keynesian money multiplier can estimate the effects of a monetary policy even in a stochastic environment and how a higher degree of financial fragility in the system raises the level of uncertainty about final effects, as stated by specialized literature.

8.1   A traditional rule of behavior for the central bank

The mechanism of the system is now modified by the introduction of a policy maker. The aggregate production is no longer left free to fluctuate but is linked to a predefined set of parameters. I refer here to the central bank as policy maker since the model does not consider public sector o public expense, focusing the attention on business fluctuations generated by firms' interaction. Furthermore, the only possible control variable for the policy maker can be identified in the interest rate that determines also the cost of capital for firms. The structure of nominal values is altered due to the presence of inflation. The reference price of the economy varies each instant by an inflation factor $\pi(t) \geq 0$ so that:

$$P(t') = (1 + \pi(t))P(t)$$

An assumption, that was implicit in the previous formulation of the model, about the capital goods market is that it is a perfect competition market.

A single firm cannot modify the price by varying the quantity bought. But in the aggregate the price of capital is determined by the intersection of the negative sloped aggregate demand and the positive sloped supply curve, therefore the central bank cannot lower indefinitely the interest rate. As highlighted below, in section 9.1, the level of interest rate determines the long path equilibrium level of N^1 and N^0 and then of production. But how does the monetary authority decide this level of interest rate? And, above all, is it really possible to control the final level of output?

Supposing that the central bank knows in advance this level of equilibrium for production and, in order to avoid fluctuations of the output around this level and to stabilize prices, adopts a Taylor rule (Taylor, 1993, as derived in Adema and Sterken, 2005):

$$r(t) = r^* + \theta_\pi(\pi(t) - \pi^*) + \theta_Y(Y(t) - Y^*) \tag{8.1}$$

r, in this hypothesis, changes over time. Parameters $\theta_{(.)}$ are decided by the central banker and quantify his sensitivity to the variables of interest, π, the inflation, and Y, the aggregate production[1]. The apex $*$ indicates the targeted optimal value. As regards output, since N is fixed, the efficacy of a stabilization policy depends on its ability to modify the probability for n^1 and n^0. In other words, also for the policy maker, the objective turns out to be the statistical equilibrium and to reach it he can handle the interest rate r. It has to be stressed that, in this perspective, the central banker focuses his attention only on the aggregate production Y, missing the fact that economy is composed by two different types of firms. The rule of behaviour of the policy implies perfect foresight (the central bank knows in advance the level of equilibrium for aggregate production Y^e) and considers the macro-variables of interest rate and aggregate production as key indicators.

In order to evaluate the effects of this kind of policy one has to measure its impact on the instant probability, since, in this way, the relative weight of n^1 and n^0, and therefore the level of output, may be modified. Moreover, influencing the probability of being in one state rather than another, the central bank could lower the degree of financial fragility in the economy. Equation (8.1) turns out to be a new information that we can incorporate in our informative set. Thus, the rule of behaviour of the central bank is embodied in the MaxEnt problem, substituting equation (8.1) in the last of the (6.2). In this way it is possible to quantify the actual effect of monetary policy on the system just by evaluating its impact on the parameter β. Indeed, as shown above, β influences both statical probability and stochastic

---

[1]To respect the Taylor principle θ_π should be ≥ 1, since the 8.1 is specified in nominal terms.

dynamics of the system and, above all, is a determinant of the steady state equilibrium level of production. The second constraint then becomes:

$$N^1 y^1(t) + N^0 y^0(t) = \frac{r(t) - r^* - \theta_\pi(\pi(t) - \pi^*)}{\theta_Y} + Y^*$$

The Lagrangean of the MaxEnt problem becomes:

$$\ell = -N^1 log(N^1) - N^0 log(N^0) - \delta_1(t)N + \delta_1(t)N^1 + \delta_1(t)N^0 + \\ + \delta_2(t)N^1 y^1(t) + \delta_2(t)N^0 y^0(t) - \delta_2(t)\frac{r(t)-r^*-\theta_\pi(\pi(t)-\pi^*)}{\theta_Y} - \delta_2(t)Y^*$$

Equating the first derivatives to 0:

$$\begin{cases} -log(N^1) - 1 + \delta_1(t) + \delta_2(t)y^1(t) = 0 \\ -log(N^0) - 1 + \delta_1(t) + \delta_2(t)y^0(t) = 0 \\ N = N^1 + N^0 \\ N^1 y^1(t) + N^0 y^0(t) = \frac{r(t)-r^*-\theta_\pi(\pi(t)-\pi^*)}{\theta_Y} + Y^* \end{cases}$$

substituting $\delta_1(t) = 1 - \alpha(t)$ and $\delta_2(t) = -\beta(t)$ and developing as done in the general case, the fourth constraint becomes:

$$\sum_{j=0}^{1} e^{-\beta(t)y^j(t)} \left\{ y^j(t) - \left[\frac{r(t) - r^* - \theta_\pi(\pi(t) - \pi^*)}{\theta_Y N} + y^* \right] \right\} = 0$$

leading to the following equation for β:

$$\beta(t) = \ln \left[-\frac{y^1(t)\theta_Y N - r(t) - r^* - \theta_\pi(\pi(t) - \pi^*) - y^*\theta_Y N}{y^0(t)\theta_Y N - r(t) - r^* - \theta_\pi(\pi(t) - \pi^*) - y^*\theta_Y N} \right] \times \\ (y^1(t) - y^0(t))^{-1}$$

Multiplying and dividing the logarithm's argument by $-N$ and considering that $r(t) - r^* - \theta_\pi(\pi(t) - \pi^*) = \theta_Y(Y(t) - Y^*)$:

$$\beta(t) = \ln \left[-\frac{\theta_Y N(y^1(t) - y^*) - \theta_Y(Y(t) - Y^*)}{\theta_Y N(y^0(t) - y^*) - \theta_Y(Y(t) - Y^*)} \right] (y^1(t) - y^0(t))^{-1} \quad (8.2)$$

After simple algebraic operations, we obtain again the (6.6).

Adopting a traditional monetary rule, which considers only aggregate values leaving no room for heterogeneity of agents, economic policy becomes completely ineffective in modifying stochastic processes of the system and, then, in stabilizing aggregate production, since the probabilities of firms of being in one state or in another remain unchanged. Even in a simple model in which heterogeneity of firms is not neglected (even though in a weak form, since firms are divided in only two groups) a monetary policy inspired by a traditional approach has actually a reduced (or no) power in pursuing its targets.

As shown in the previous chapter, parameter β also affects the dynamics of the system. Therefore, without modifying it, the central bank cannot interfere with the stochastic dynamics. Considering a deterministic transmission of interest effects, it is in the end not possible to determine the effective level of production. It is the effect of the unpredictable stochastic dynamics of the system, illustrated by figure 9.6 and 9.7 in the next chapter, for which it is not useful to treat interest rate as a control variable in order to modify aggregate output.

8.2 A New-Keynesian perspective

As should be evident, if the policy maker would target only one of the two fractions of production (therefore setting as objective Y^0 or Y^1), it could actually modify the equation of the quantum anomalies and the estimation of β and, in this way, the static probabilities for firms to be in one state or another. But, any way, in order to apply a Taylor rule, he should know the optimal level of Y^0 or Y^1. In a stochastic environment (as well as in the real world) the knowledge of an agent hardly goes so far.

Delli Gatti (1999) suggests an alternative approach for studying the transmission mechanism of monetary impulses, taking into account the structural differences among firms. Adopting a New Keynesian perspective, the quantity of money (M) is considered as the control variable for the central bank. Therefore the potential effect of a variation in M is evaluated by its impact on the aggregate production, through the money multiplier mechanism. In this approach, the central bank targets the financial fragility of heavily indebted firms as the potential critical factor for a stabilization policy. Recent literature has stressed the potential risk of firms' excessive debt level for instability of the financial market and at macroeconomic level (Bean, 2004). The consequent tightening and insecurity in the credit market makes it more difficult for the authority to achieve an effectual policy. The structure of the economy and the difference in firms' financial state critically influence the impact of a monetary policy. Some basic facts are worth considering:

- monetary policy has an asymmetric effect on output: the aggregate production function is discontinuous, since the level of output is the same for all firms in state y, whilst it progressively reduces as the net worth decreases (and therefore the probability of bankruptcy increases) for firms in state x. Then, an intervention of the central bank can burst a non-linear effect since it may cause a transition for firms from one state to another. The actual come out can be estimated only taking into account the relative proportion of firms in the two states and their relative probabilities of transition;

- as detailed in the previous section, the only way to stabilize production is to reach statistical equilibrium; then, any attempt to modify the dynamics of the system has to be evaluated in probabilistic terms;

- firms are heterogeneous and, therefore, the same variations in the general financial situation do not imply an equal reaction for all firms.

Put in these terms, the effect of an economic policy has to be measured in terms of its impact at micro level and, more specifically, on its capacity to modify firms transition probabilities. Since firms are clustered on the basis of their net worth level and that the probability of bankruptcy is an inverse function of it, the impact of the multiplier on the production can be expressed in terms of the net worth variation, as in Delli Gatti (1999).

Formally, the expected total variation in output due to an infinitesimal variation in the supply of money is:

$$\frac{dY}{dM} = \left(N^1 \frac{dY}{d\mu} \frac{d\mu}{da} \frac{da}{dM} \right) \left[N^0 \left(\frac{d\zeta}{da} \frac{da}{dM} \right) - N^1 \left(\frac{d\iota}{da} \frac{da}{dM} \right) \right] \qquad (8.3)$$

The value of the multiplier for aggregate production is:

$$\frac{\partial Y}{\partial M} = \frac{N^1 2c}{(r + 2c\mu)^3} q^1 \frac{\partial a}{\partial M}$$

The effect of a modification in the supply of money on total output is not limited to the change in the relative probability of bankruptcy, but it also involves the induced transitions of state of firms, since a firm that alters its state also modifies its optimal level of production. Operating in a stochastic environment, the only way we have to measure this second-type effect is to quantify the relative adjustments in the transition rates, by means of first partial derivatives:

$$\frac{\partial \zeta}{\partial M} = -\frac{\partial a}{\partial M} (r + 2c\mu)^{-1}$$
$$\frac{\partial \iota}{\partial M} = \frac{\partial a}{\partial M} (r)^{-1}$$

After some simple calculations we are able to express the effect of the multiplier as:

$$
\begin{aligned}
\frac{\partial Y}{\partial M} &= \left(\frac{\partial a}{\partial M}\right)^2 \frac{N^1 2c}{(r + 2c\mu)^4} \left(\frac{N^0}{r} + \frac{N^1}{r + 2c\mu}\right) = \\
&= \left(\frac{\partial a}{\partial M}\right)^2 \frac{N^1 2c}{(r + 2c\mu)^4}
\end{aligned}
\tag{8.4}
$$

As in Delli Gatti (1999) the overall effect depends on the number of firms with the worst financial position (N^1 in the present work). In the limit case of $N^1 = 0$ the multiplier goes to 0 as well (the effect on the production function through firms' equities becomes null). The remarkable difference with the cited model is that, working in a stochastic environment, we cannot conclude that the multiplier strengthens as N^1 increases, but, rather, that the final equilibrium distribution of the effects becomes increasingly sensitive to M. This, on the other hand, makes the system more unstable since, in equation (8.4), N^1 is a multiplicative factor for stochastic variable μ, and widens the range of possible outcomes of the policy. In other words, a growing financial fragility in the system increases the level of uncertainty about actual policy results, consistently with Bean (2004).

Both models verify that, in the presence of heterogeneity (of size or financial condition) and the consequent uncertainty, a policy maker should use proper analytical tools to take any kind of action. The use of representative agent hypothesis and a determistic representation of the transmission mechanisms may determine a misperception (due to the exclusive use of macro indicators) and, finally, an undesired effect on the system.

Part III

Empirical evidence

Chapter 9

Application to data

This chapter reports the results of numerical applications of the model introduced in part II. The first section regards the outputs of simulations, performed by means of Matlab software. In section 9.2 the model is applied on empirical data, at micro and macro level, for different countries.

9.1 Simulations results

The stochastic structure defined in chapter 5 represents a suitable analytical instrument to study different possible scenarios. In order to illustrate in detail the mechanism of the system, its capacity to return useful information and check the sensitivity of the results to (static and dynamic) variations in the parameters, I performed some simulations by means of Matlab software.

In order to get a treatable analytic approximation of mean-field interaction, equity ratios for firms in the two state have been set to:

$$a^0(t) = \bar{a} + \epsilon\sigma(t)$$
$$a^1(t) = \bar{a} - \epsilon\sigma(t)$$
(9.1)

where $0 < \epsilon \leq 1$ and $\sigma(t)$ is the standard deviation in equity ratio distribution at time t. The first random sample of firms equity ratio is drawn from an exponential distribution[1] with a given σ.

Numerical simulations provide some insights about the dynamics of the percentage of firms in state x, n^1, toward its equilibrium level, m^* (equation (6.15)) and its dependence upon model's parameters. The first series of graphics shows the effects of changes in significant parameters on the velocity of the adjustment, on the final level of equilibrium and on amplitudes of

---

[1]The empirical distribution of equity ratio is well approximated by an exponential, as assessed in section 9.2.1.

fluctuations. Simulations illustrate how the level of long term equilibrium is determined by structural variables and by agents' distribution. In the short term, uncertainty negatively affects economy's performances, unless accompanied by a generally sounder financial situation.

As already demonstrated by means of equation (6.17), the dynamics is convergent, whatever the initial level of m. Figure 9.1 displays that the velocity of adjustment is heavily dependent on initial conditions. Systems in better initial financial situations (lower level of $m(0)$) take a longer time to reach the long run steady state, continuing then to maintain a higher level of production respect to an economy that reaches the equilibrium path starting from a higher $m(0)$.

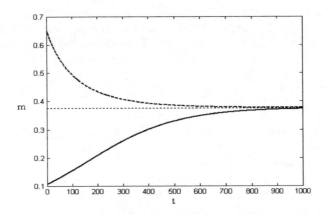

Figure 9.1 – Drift dynamics for different initial percentages of firms in state x: $m(0) = 0.1$ (continuous line) and $m(0) = 0.7$ (dashed line).

The final equilibrium percentage of "bad" firms is modified by interest rate and by standard deviation of equity ratio distribution. As expected, a higher level of interest implies a higher value of m^*, since it raises the probability of entering in x, ζ, and lowers the probability of exit, ι (figure 9.2). At the same time, for a fixed level of interest rate, a low standard deviation of equity ratio determines a higher final m^* (figure 9.3). Computationally, this result is due to equations (5.17), (5.18) and (9.1): an increment in $\sigma(t)$ determines decrease in both the transition rates ζ and ι, but the lowering in ζ is faster and, by means of equation (6.15), origins a lower m^*.

Interest rate level also influences the amplitude of the oscillations, as depicted in figure 9.4. Precisely, a lower interest rate causes the range of

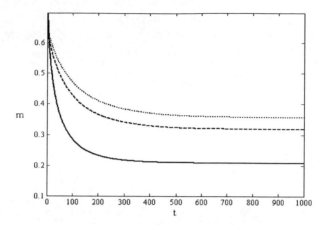

Figure 9.2 – Drift dynamics for interest rate equal to 0.05 (solid line), 0.1 (dashed line), 0.15 (points).

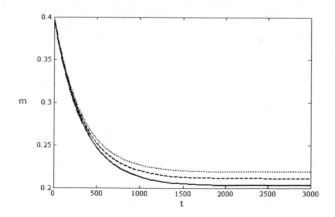

Figure 9.3 – Drift dynamics for equity ratio distribution's standard deviation equal to 0.4 (solid line), 0.3 (dashed line) and 0.2 (points).

output fluctuations to reduce. This is due to the lower number of firms that occupy state x. As shown in section 8.2, the level of n^1 constitutes a factor of uncertainty since the optimal level of output for "bad" firms is subject to oscillations, due to the presence in their object function of the probability of demise $\mu(t)$. Then, a higher percentage of "bad" firms implies a higher volatility in output.

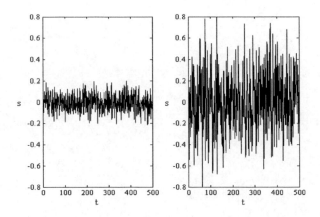

Figure 9.4 – Spread for $r = 0.05$ (left panel) and $r = 0.10$ (right panel).

Figure 9.5 depicts trend and fluctuations components of the specular transitional dynamics of the fraction of firms m and of the total production Y.

Figure 9.6 reports the bifurcation diagram. For higher values of r, some ceiling effects (on the transition rates) operate and, therefore, the dynamics cannot be explosive. For realistic value of interest rate (below a level around 10%), the system generates a chaotic dynamics. Figure 9.7 shows the detail of bifurcation diagrams for two different initial levels of m, displaying that a higher starting value of $m(0)$ will always imply, also for a level of interest rate close to 0, a higher m. Moreover, as can be noted, higher value of $m(0)$ will also generate a wider range of variation in the chaos region. Therefore, as demonstrated in section 8.2, in analyzing the chance of monetary policy to influence the system's behavior, uncertainty about the effect of a variation in the interest rate comes out be positively correlated to the level of financial fragility of the economy.

In order to deepen the system's behavior over the range of variation of equity ratio's standard deviation, σ, I perform another series of simulations, focusing the attention on how parameter β and aggregate production, Y,

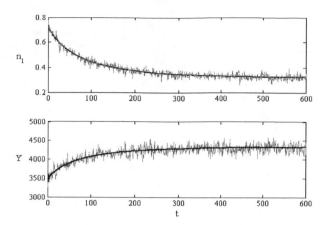

Figure 9.5 – Dynamics for logistic drift and endogenous spread for n_1 (upper panel) and for aggregate production (lower panel).

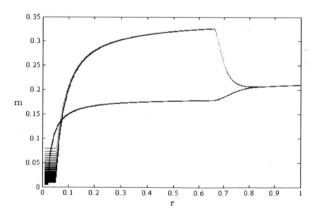

Figure 9.6 – Bifurcation diagram for m as a function of the interest rate r.

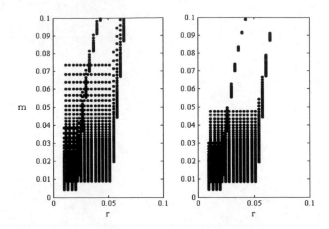

Figure 9.7 – Bifurcation diagram for $m(0) = 0.4$ *(left panel) and for* $m(0) = 0.1$
(right panel).

respond to variations in the level of dispersion in equity ratios' distribution.
As displayed in figure 9.8 the value of β remains around 0, signifying remarkable uncertainty: firms occupy about in the same proportion the two
states. As one could expect, as σ rises, the range of variation of β enlarges,
or, in other words, the uncertainty and the sensitivity of "bad" firms to the
probability of failure μ increase as well. Keeping in mind equation (6.8), the
variability in the system appears to be due to the proportion of firms in the
two states (determined by σ) and to their expected bankruptcy costs in a
variable proportion.

Figure 9.9 shows the sensitivity of aggregate production (Y) to the standard deviation of equity ratio. Aggregate output appears to be negatively
correlated with large values of β and positively with σ. For values of σ close
to 0 or to 1, some frontier-effects might operate due to the normalization rules
for transition rates (5.12). The negative effect of large values of β might be
due to an attenuation of the positive effect of an increase in n^0 caused by
a lowering value of the overall mean equity ratio, since, in an exponential
distribution, standard deviation and mean have the same value.

9.2   Application to real data

In this section the model is simulated using real micro and macro data, in
order to test its capacity to return useful information about national systems'

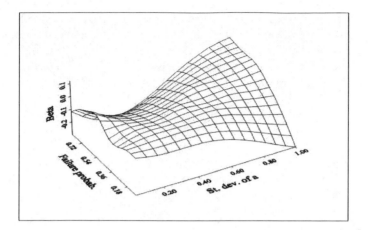

Figure 9.8 – Values of parameter β as a function of the probability of bankruptcy μ and the standard deviation of equity ratio distribution. The values of the parameters are: c=0.3; r=0.15; N=100; ε=0.7; initial value of q¹=4; range of variation of ũ(t) = [0.05; 1.95].

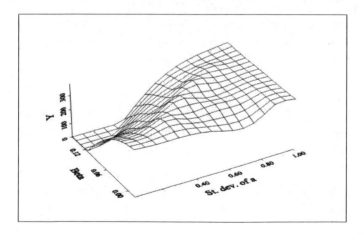

Figure 9.9 – Levels of aggregate production as a function of β and standard deviation of equity ratio distribution. Same levels of the parameters.

financial structures in such detail that cannot be inferred by a simple analysis of aggregate variables and indicators. In the matter of financial soundness of firms four very different national contexts are taken into examination, and, namely, two samples of non listed firms and two samples of listed ones: great firms for Italy and France and listed Australian and USA firms[2]. Data for Italy and France are taken from the commercial data set Amadeus, containing firm level data for great firms[3] in the period 1992-2005. For Australia the reference data set is Aspect Fin Analysis that provides balance sheet data for companies listed on the Australian Stock Exchange (ASX) in the period 1989-2006. USA listed firms' data are provided by Compustat dataset, that covers the period 1962-1999. Macro data are provided, for all countries, by World Bank Development Indicators 2006. Lending interest rates (at net of inflation, measured by GDP deflator) are used as proxies for real interest rates. Average added value (operating profit for Australia and USA) of the included firms is used as a proxy of aggregate production, in order to avoid distortions due to the variability, from one year to another, in the number of firms included in the sample. The use of GDP as a proxy of Y would include variations due to externalities that may lead to ambiguous results[4].

In order to determine a critical threshold for firms that should indicate a possible risk of demise ($\bar{a}$) I make use of the existing evidence on the matter. Figure 9.10 shows the trend of average equity ratio of failed firms as bankruptcy approaches for two of the countries in exam. An acceptable approximation to the critical threshold is an equity ratio equal to 0.1, that corresponds to the value at about two years before demise in France. Values of all the other parameters are kept unchanged from the simulations presented in previous section (specifically: c=0.3; ϵ=0.9 and range of variation of $\tilde{u}(t) = [0.05; 1.95]$).

9.2.1   Equity ratio distribution and detailed balance

Preliminarily, the distribution of equity ratio within nations has been investigated. In all cases exponential distribution returns a very good fit (see as a matter of example figure 9.11).

---

[2]The risk for listed firms is not mainly the bankruptcy but, rather, the exclusion from the stock exchange. According to Karl Popper's principle of falsifiability, a theory should be considered scientific if and only if it is falsifiable (Popper, 2002). Therefore, the check for an eventual different behavior of listed and non listed firms constitutes an interesting test for the reliability of the model.

[3]Namely firms that satisfy one of these criteria: at least 15 million Euros of operating revenues, or 30 million Euros in assets or 200 employees.

[4]As regards Australian firms, the results are comparable using average added value or GDP.

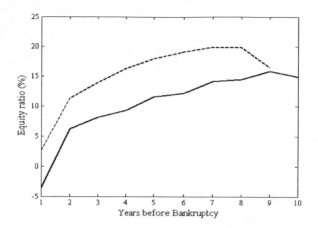

Figure 9.10 – Average equity ratio for bankrupted firms in Italy (continuous line) and France (dashed line). Source: elaboration on Amadeus data.

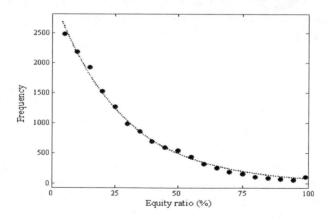

Figure 9.11 – Italy 2004. Distribution of equity ratio with exponential fitting line. Source: elaboration on Amadeus data.

The comparison among the percentages of firms below the threshold records significant differences between the countries (see table 9.1, that displays only the years for which data are available for at least two countries). While a lower rate with reference to Australia and USA could be expected (since the relative data sets include only listed companies), the statistic for Italy appears to be remarkably high. The evidence comes out to be alarming if one considers that, fixing an equity ratio's threshold of 0.2, the percentage of firms with equity ratio below $\bar{a}$ rises to over 50% for all the period.

Year	Australia	France	Italy	USA
1989	0.025547445	Na	Na	0.111696523
1990	0.020033389	Na	Na	0.109258979
1991	0.0352	Na	Na	0.108251996
1992	0.052478134	0.078947368	0.271159875	0.093146853
1993	0.049872123	0.111111111	0.296766744	0.130298651
1994	0.030567686	0.112582781	0.270308789	0.123462467
1995	0.026342452	0.087912088	0.232502612	0.109608159
1996	0.02575897	0.137648505	0.259994585	0.105164515
1997	0.021276596	0.140045795	0.274879548	0.100617351
1998	0.023237179	0.139499797	0.270723604	0.103772516
1999	0.026912181	0.146634749	0.27543036	0.105332111
2000	0.024016011	0.148102438	0.273888363	Na
2001	0.02277294	0.143432557	0.27724359	Na
2002	0.023399862	0.141040462	0.2745338	Na
2003	0.023411371	0.1391348	0.265893361	Na
2004	0.032756489	0.13725947	0.257509176	Na
2005	0.019562716	0.117417919	0.244474515	Na

TABLE 9.1 – *Relative proportion of firms whose equity ratio is below 0.1. Source: elaboration on Amadeus data. Na: data not available.*

An interesting feature that emerges from the analysis is the constantly increasing trend of "bad" firms in the US stock exchange, displayed in figure 9.9. The percentage passed from negligible values during a first period to a level always above 10% in recent times and, as evident from the graph, the changes do not seem to depend on particularly low interest rates.

The impression that the phenomena is somehow structural is confirmed by dynamic analysis, performed in order to test the eventual steady state condition (i. e. the statistical equilibrium, as exposed in previous chapters) of the systems in exam. The study should enlighten on whether the examined economies are in a transition phase or along their steady state path (as depicted in figure 9.5). The existence of the detailed balance is verified by

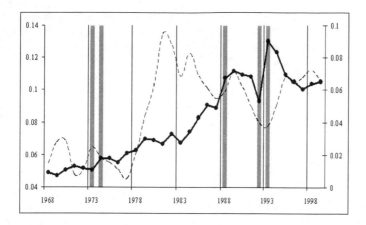

Figure 9.12 – Trends of fraction of US listed firms with equity ratio below 0.1 (continuous line with marker points) and real lending interest rate (dashed line, right axis). Grey areas: rejection of detailed balance. Source: elaboration on Compustat and World Bank data.

means of the two dimensional Kolmogorov Smirnov test (Fasano and Franceschini, 1987), following the algorithm proposed in Smallwood (1996). The null hypothesis (detailed balance condition holds) is accepted in all countries for almost all the periods in exam and, in particular, for all the years in which there is a consistent number of observations. Interestingly, in USA, null hypothesis is rejected only in 1973 and 1974 (immediately after the first oil crisis), in 1988 (a year after the "black Monday" in Wall Street) and during turbulences in international financial markets in 1992-93. Also in the last covered years null hypothesis is rejected but the test is not significant for the low number of observations. Therefore, it seems that, for most of the period, the shown upward trend in n^1 is compatible with statistical steady state condition and only abrupt changes in general financial conditions determined significant modifications on equity ratios distribution, and, then, the violation of detailed balance condition.

9.2.2 Financial fragility and performances

In this paragraph, the relationship between the degree of financial fragility of firms and their performances, in terms of produced added value, is investigated. Financial fragility is measured, in a twofold perspective, by the proportion of firms with equity ratio below 0.1 (a threshold exogenously determined) and probability of failure (obtained using real data to calculate the outcome of equation (5.13)). Consistently with the model, the dependence of firms' results on the average financial soundness of the system is more evident for Italy and France, where non listed firms are included in the data set (figures 9.13 and 9.14), than for Australia and USA, whose data regard only publicly traded firms (figures 9.15 and 9.16). The negative impact of both variables is particularly remarkable for France, while for Italy the negative effect of risk of demise appears to be more pronounced, as one could expect given the observed high levels of n^1.

This empirical fact also shows how the rationing on equity market negatively affects firms performances, consistently with model's results. As regards listed firms a negative correlation among financial indicators and performances comes into sight in a clearer way for Australian companies. For USA even a positive relationship between average profits and percentage of "bad" firms seems to emerge. It has to be considered, in this case, that, given the long time span of observation, there is a physiological increase in productivity jointly with a constant increase in n^1 percentage, as verified in figure 9.12.

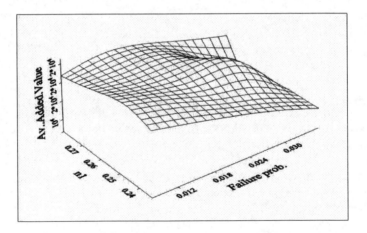

Figure 9.13 – Italy 1992-2005. Average added value as a function of the probability of failure and of proportion of firms with equity ratio below 0.1.

9.2.3 The index β and the aggregate output

Italian firms display a lower standard deviation, compared to the other three countries, with values of the equity relatively concentrated. As mentioned,

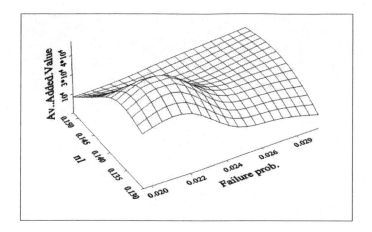

Figure 9.14 – France 1992-2005. Average added value as a function of the probability of failure and of proportion of firms with equity ratio below 0.1.

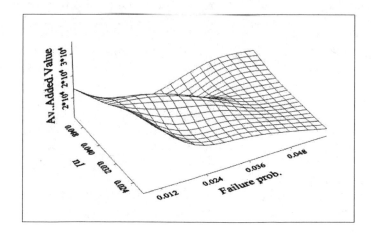

Figure 9.15 – Australia 1989-2005. Average added value as a function of the probability of failure and of proportion of firms with equity ratio below 0.1.

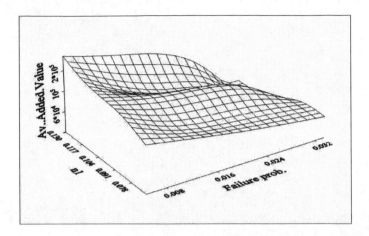

Figure 9.16 – USA 1962-1999. Average added value as a function of the probability of failure and of proportion of firms with equity ratio below 0.1.

with the slightly higher threshold of $\bar{a} = 0.2$, n^1 appears to be always bigger than n^0 with a value of β constantly negative. For $\bar{a} = 0.1$, figure 9.17 displays a negative relationship between average added value and dispersion in equity distribution, and the relationship holds for different values of β. Such an evidence suggests that, since a bigger standard deviation implies a more difficult switching of state and, in a context where a relevant quote of firms is in a difficult financial situation, it may determine a negative effect on the aggregate.

Regarding France, figure 9.18 displays an opposite behavior: the growth of σ is accompanied by an increase in the added value. As equity ratio's distribution is exponential, a higher standard deviation implies a higher mean. In other words, the bigger level of production is accompanied by an average improvement in the financial situation of firms. This effect seems to be enforced by a greater dispersion, since the relationship appears to be clearer for values of β closer to 0.

A third different situation is shown in figure 9.19 for what concerns Australia. The linkages among variables are not evident and only at the lowest level of equity's standard deviation a positive correlation among β and average added value seems to emerge, probably for a low mean in the overall equity that makes aggregate production more sensitive to a variation in the proportion among n^1 and n^0 or to a variation in r.

An analogous result (with different sign) is found for USA (figure 9.20) for which no correlation among variables appear, except for a few points (that

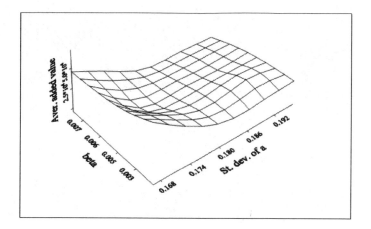

Figure 9.17 – Italy 1992-2005. Average added value for firms included in Amadeus sample as a function of β and standard deviation of equity ratio distribution.

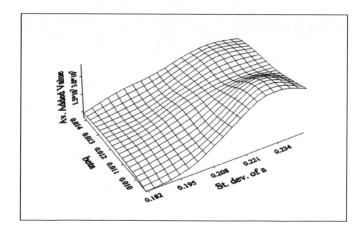

Figure 9.18 – France 1992-2005. Average added value for firms included in Amadeus sample as a function of β and standard deviation of equity ratio distribution.

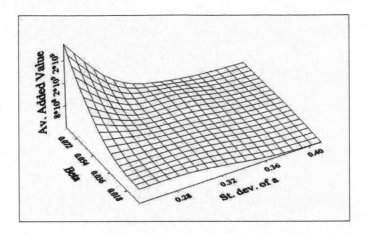

Figure 9.19 – Australia 1989-2005. Average operating profit for firms listed on the Australian Stock Exchange (ASX) as a function of β and standard deviation of equity ratio distribution.

correspond to the last years of the sample, from 1994 to 1999) for which a positive relationship among dispersion and output, only for the highest values of β, seems to emerge. As, one could expect, a stricter correlation among financial variables and indexes and firms' performances appears only for the two countries whose data sets include non listed firms. Such evidence can be regarded as a consequence of the rule of inclusion in the sample (since listed firms are - or should be - financially sounder) but, mainly, on the possibility for listed firms to access to equity market, keeping in mind that one of the base hypothesis of the theoretical model is that firms are fully rationed on the equity market. This stylized fact is further investigated in chapter 11.

As regards cyclicity, for Australia, France and Italy, standard deviation of equity ratios displays a relevant negative correlation with interest rates (see figure 9.21 for Italy). Inverse relationship that appears particularly evident for Italy and France, where a constant increase in dispersion of distribution accompanied the diminishing trend observed for interest rates. For USA it appears clearer only in the second half of the observation period. In all cases the behavior of β appears to be negatively correlated with standard deviation and strictly positively with interest rates. This last result reveals that an increase in the use of debt as source of finance for firms cannot be simply explained by a reduction in interest rates. This aspect is further documented in section 10.3.

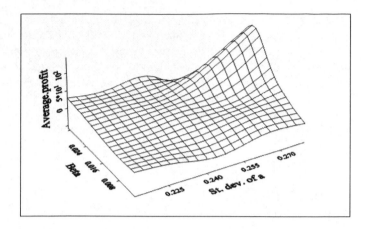

Figure 9.20 – USA 1962-1999. Average operating profit (in thousands dollars) for firms listed on the New York Stock Exchange as a function of β and standard deviation of equity ratio distribution.

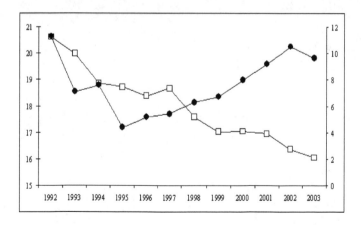

Figure 9.21 – Italy 1992-2004. Standard deviation of equity ratio distribution (full circles) and real lending interest rates (empty squares, left axis), in percentage. Source: elaboration on Amadeus and World Bank data.

Finally, it has to be remarked that, paying attention only to the macro variables or to aggregate indicators, policy makers could obtain ambiguous, or even erroneous, indications. This analysis showed how macro variations have to be interpreted in the light of the underlying distributions of agents. The consideration of the statistical features of the system, jointly with the use of the parameter β, here introduced, permits to enlighten what happens below the surface and could help the decision making process, for example in the matter of interest rates and support to indebted firms.

Chapter 10

Stylized facts of business cycles

Good economists know the universal truths
and can look beyond the array of facts
and details that obscure these truths.
(Stiglitz, 2000)

In this chapter I report some stylized facts regarding business cycle phases. The reason to devote a separate chapter to the macroeconomic evidence consists, mainly, in presenting, in greater detail, some of the facts already mentioned or only sketched in previous chapters and, above all, in showing the importance of firms' financial variables in determining macro evidence. In the following (consistently with the approach adopted in this work) we summarize existing evidence and present some new facts about statistical emergences at macro level, showing how business cycle variables are distributed and suggesting some considerations that can be inferred from this evidence. In what follows I adopt the NBER definition for business cycles (Burns and Mitchell, 1946). Then, single year changes are simply referred to as positive and negative variations, quantified by the percentage variation in GDP. Cumulative variations are defined as expansions and contractions and expressed in terms of deviations from an estimated GDP trend or potential (Zarnowitz, 1992), identified by means of Hodrick and Prescott low-pass filter (Hodrick and Prescott, 1998). The first analysis regards fluctuations' basic features, such as duration and intensity of phases. Then in section 10.3, after a brief review of existing empirical literature, the relationship among financial fragility and macroeconomic indexes is evidenced and investigated.

10.1   Durations

As briefly discussed in section 6.2.2, statistical distribution of durations for business cycle phases gave rise in recent years to different analysis, whose results appear to be partially contradictory. For Ormerod and Mounfield (2001) durations of contractions follow a power law distribution while, according to Wright (2003), data are better fitted by an exponential law. For Ausloos et al. (2004) the duration of recession phases is power law distributed, while expansions' durations distribution is exponential[1]. All these studies took the year as a unit of measure of time and, then, met the limit of the very scarce number of observations to perform a satisfactory distribution fitting.

In order to avoid this problem, I used monthly data from December 1854 to November 2001, extracted by a public NBER data set, in order to check the fit of power law distribution on the data. The calculation of parameters is performed by means of the method introduced in Clementi et al. (2006). This technique relies on the popular Hill's maximum likelihood estimator (Hill, 1975) for the tail index and a subsample semi-parametric bootstrap algorithm for datadriven selection of the number of observations located in the tail of the distribution.

Pareto distribution has probability function:

$$p(x) \propto x_0/x^\alpha$$

being x_0 the minimum value of the observations included in the tail of the distribution. A correct estimation of x_0 assumes particular importance for the maximum likelihood estimation of the shape parameter α. As far as the author knows, this is the only datadriven technique for calculation of the Pareto threshold parameter. The method permits to avoid the problems implied by the OLS estimator (Newman, 2005), used in the cited works.

Results are shown in figures 10.1 and 10.2. The vertical dashed line identifies the tail of distribution and, therefore, defines in statistical terms the minimum level of maturity to define a cycle's phase, to coin the words of Diebold and Rudebusch (1990). The optimal threshold estimations are 21 months for expansions and 7 for recessions, while the shape parameters are, respectively, 1.6881 and 1.2949. These coefficients come out to be different from the estimations reported in Ormerod and Mounfield (2001) since the data set, the time sampling frequency and the estimation methods are different.

---

[1]Quite surprisingly, Ausloos et al. (2004) do not apply any filter to the data in order to clean them by the trend. We cite their work here uniquely for the completeness of the review.

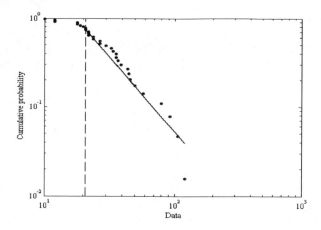

Figure 10.1 – Power law fit for duration of USA expansions (in months). Source: elaboration on NBER data.

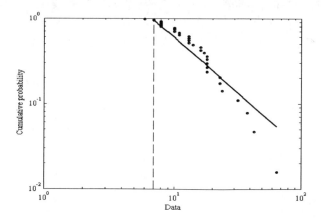

Figure 10.2 – Power law fit for duration of USA recessions (in months). Source: elaboration on NBER data.

10.2 Amplitude

In this section variations in aggregate production during business cycles phases are analyzed. I take into exam different measure of change as annual variations, cumulative variations (proper expansions and recessions) and *steepness*, defined by Di Guilmi et al. (2005) as the ratio among the absolute value of the cumulative percentage points of peak-to-trough and trough-to-peak output gap for recessions and expansions, respectively, to the time duration of the phase (expressed in number of periods). In geometrical terms this measure corresponds to the slope of the hypotenuse of the triangle approximation to the cumulative movement during a business cycle phase, as discussed in Harding and Pagan (2002).

The data set is provided by the International Monetary Fund (2002) and covers a time span of 120 years (from 1881 to 2000) for 16 industrialized countries[2]. The two world war periods have been excluded. Positive and negative variations are studied separately.

Annual rates of variation of GDP, both positive and negative, follow a two parameters Weibull distribution, as shown in figures 10.3 and 10.4. The formulation of probability function is:

$$p(x) = ba^{-b}x^{b-1}e^{-\left(\frac{x}{a}\right)^{-b}} \tag{10.1}$$

As figures show, fit appears to be excellent, particularly for expansions.

As regards fit and parameters' estimations for cumulative variations and steepness, I refer here to Di Guilmi et al. (2004, 2005). Expansion and recession rates (figures 10.5 and 10.6) and steepness (figures 10.7 and 10.8) distributions are well approximated by a Weibull probability function. The maximum likelihood estimations, at 95 % confidence level, for all distributions, are reported in table 10.1[3].

---

[2]Namely: Australia, Canada, Denmark, Finland, France, Germany, Italy, Japan, The Netherlands, Norway, Portugal, Spain, Sweden, Switzerland, UK and USA.

[3]Di Guilmi et al. (2005) report estimations for two different hazard models, that correspond to the two and three parameters formulation of Weibull density. In order to take comparable values, estimations here reported correspond to the model of Lancaster (1979), that employs a two parameters Weibull function.

Variable	Phase	a	b
Single year variation	Expansion	2.9900 ± 0.1887	1.0875 ± 0.0553
Single year variation	Recession	2.9390 ± 0.2058	0.9883 ± 0.0490
Cumulative variation[+]	Expansion	0.1340 ± 0.0387	1.3445 ± 0.1405
Cumulative variation[+]	Recession	0.146 ± 0.0423	1.2383 ± 0.1228
Steepness[++]	Expansion	0.0387 ± 0.0024	1.0143 ± 0.0384
Steepness[++]	Recession	0.0401 ± 0.0018	1.3832 ± 0.0604

TABLE 10.1 – *Empirical estimation for Weibull distribution parameters.* [+] *source: Di Guilmi et al. (2004).* [++] *source: Di Guilmi et al. (2005).*

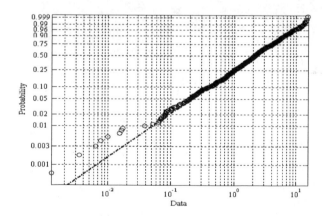

Figure 10.3 – Weibull probability plot for distribution of annual GDP positive variation. Source: elaboration on FMI data.

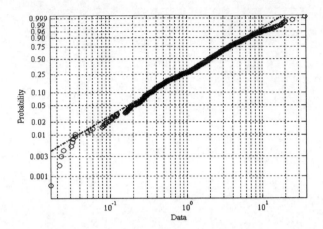

Figure 10.4 – Weibull probability plot for distribution of annual GDP negative variation. Source: elaboration on FMI data.

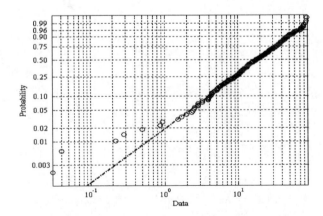

Figure 10.5 – Weibull probability plot for cumulative percentage variations in expansions. Source: Di Guilmi et al. (2004).

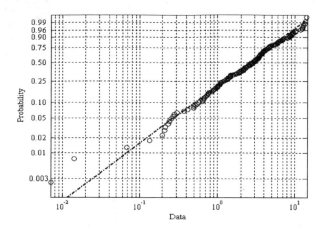

Figure 10.6 – Weibull probability plot for cumulative percentage variations in recessions. Source: Di Guilmi et al. (2004).

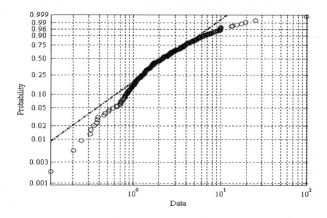

Figure 10.7 – Weibull probability plot for steepness for expansions. Source: Di Guilmi et al. (2005).

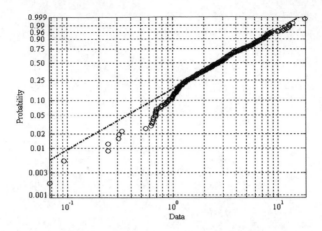

Figure 10.8 – Weibull probability plot for steepness for recessions. Source: Di Guilmi et al. (2005).

10.3   Firms' financial fragility and business cycles

Previous parts of this work highlighted the role of (the probability of) bankruptcies in determining fluctuations and, also, macro trends. Equity ratio has been indicated and adopted as reference variable to estimate the probability of failure. A practical reason for its use, among other structural and balance sheet items and indexes, is that it is available on all the firm level data sets I employ here and its calculation is unambiguous, using the same balance sheet entries in all examined countries.

Moreover, there is wide evidence about the correlation between financial soundness or distress of firms and their profitability and, in particular, about the effectiveness of equity ratio as a synthetic index, for predicting or estimating performances[4]. Figure 9.10 already documented the progressive diminishing of equity ratio for failed firms as bankruptcy approaches. Figure 10.9 provides more detail, showing, for Italy, the frequency of demises as a function of equity ratio's value recorded in the year before the demise. Actually, with a ratio over the 30% the probability of failure drops to zero. Figure 10.10 offers an idea of the direct relationship among financial soundness and performance of an economy, which is at the root of the model presented in part II. The example regards Italy and the graph exhibits mean and stan-

---

[4]This part could be put at the beginning to indicate, preliminarily, the reason of this choice. However, I prefer to group together all empirical evidence.

dard deviation of return on assets (net profit on total assets) for bins of firms identified by their equity ratio. There is a clear and almost monotonic relationship between results and quote of internal source of financing. The values for firms with equity bigger than 80% is mainly due to the limited number of observations, as testified by the variability. The statistical relationship between equity and profits is deeper investigated, at micro level, in the next chapter, by means of hazard function analysis.

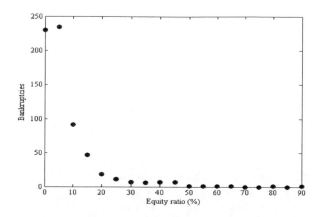

Figure 10.9 – Italy 1993-2004. Number of bankruptcies as a function of equity ratio (in %). Source: elaboration on Amadeus data.

Some empirical studies found evidence about the burden of financial position of firms in determining their bankruptcy, adopting different proxies for measuring credit risk (Altman, 1968; Fulmer et al., 1984), developing models that are usually adopted by financial institutions to decide about requests of credit. Another work of Beaver (1967), studying a cross section analysis of US firms by means of three different tests, shows how deep balance indexes worsen progressively in years before bankruptcy.

Other structural indicators have been analyzed in literature. For instance, as regards age of failed firms, Cooley and Quadrini (2001) show that young firms are more likely to exit than old ones. The use of age as a predictor for failure meets a fundamental drawback. As demonstrated by Fujiwara (2004) and Delli Gatti et al. (2004), age of bankrupted firms follows an exponential distribution. The Markov (or lack of memory) property of the exponential distribution lets us assert that the probability for a firm to exceed a certain age x is independent from whether x has already been reached or not, i.e.,

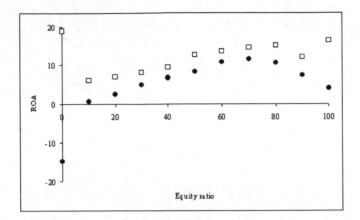

Figure 10.10 – Italy 1992-2005. Mean (circles) and standard deviation (empty squares) of return on assets conditioned on equity ratio. Source: elaboration on Amadeus data.

using only age, it is not possible to formulate a prediction about the residual life of firms. The demonstration is provided by Azlarov and Volodin (1986). Let us define the exponential distribution function as:

$$P(x) = e^{-\alpha x}$$

The probability for the random positive variable ϵ of reaching x, given that y has already been reached, is given by:

$$P(\epsilon \geq x + y | \epsilon \geq y) = \frac{P(\epsilon \geq x + y)}{P(\epsilon \geq y)}$$

The lack of memory (or Markov property) consists in the assumption that $P(\epsilon \geq x)$ is not conditioned by $P(\epsilon \geq y)$:

$$P(\epsilon \geq x + y) = P(\epsilon \geq x)\, P(\epsilon \geq y)$$
$$\Updownarrow$$
$$P(\epsilon \geq x + y | \epsilon \geq y) = P(\epsilon \geq x)$$

It is always verified for the exponential distribution since:

$$P(x + y) = e^{-\alpha(x+y)} = e^{-\alpha x} e^{-\alpha y}$$

As briefly summarized in section 2.2, the relevance for firms' financial variables in determining economic fluctuations has been widely addressed in

Keynesian and New Keynesian literature and, specifically, from a theoretical point of view, by the *Debt Deflation school* (Fischer, 1932; Kindleberger, 2005; Minsky, 1982). Among others, Dow (1998) stresses the role of debt as a factor of weakness for firms in facing negative shocks, even if he then argues that debt itself cannot be considered as a cause of downturns. Actually, it is quite intuitive that heavily borrowed firms will probably abruptly reduce their investments as they perceive a contraction in final demand, reinforcing depression[5].

At macro level, wide evidence regarding the Financial Instability Hypothesis (FIH) have been detected. In particular, limiting our review to the debt-aggregate indicators relationships, Goldfajn and Valdes (1998)[6] found positive relationships between M2/reserves ratio and the probability of a currency crisis and among the level of domestic credit, high public debt and financial deregulation with the probability of a banking system crisis. The risk of default has also been connected to the famous (and apparent) Lucas' paradox of capital flows (Lucas, 1990) that, breaking neoclassical indications, do not go from rich to poor countries, finding substantial evidence in favor of an explanation *a lá* Minsky (Reinhart and Rogoff, 2004).

FIH gained momentum in recent years, from Far East crisis, in explaining international financial disturbances (Kregel, 1998). An interpretation of the crisis based on micro-level variables had been suggested by Setser et al. (2002) and formalized in the so-called *Balance Sheet Approach*. In this perspective, economy is modeled as a system of balance sheets of all its agents, and its macro-variables are consequently determined, in particular, by the stocks of assets and liabilities. The Balance Sheet Approach has been integrated in a FIH approach by Toporowski and Cozzi (2006), in a framework that is very sympathetic with the present work, putting emphasis on the transmission of financial instability from non-financial firms to the financial sector, where it emerges and starts domino effects. Minsky's work has been inspiring also at a predicting level. In particular, Kaminsky, in a series of papers[7] developed a system of anticipating indicator (*early warning indicators*) that estimates the probability for a system to be hit by a financial distress in the two following years, given the present level of the indicators.

Nowadays, the (absolute and relative) dimensions that levels of private and business debt are reaching in capitalistic economies amplifies the rele-

---

[5]Dow, in the same work, counterarguments that, in alternative, firms may emit equities instead of borrowing from external sources. On the other side, following the theory of hierarchy of financial sources, the emission of new equities is not the first choice for firms, that prefer to borrow (Gertler, 1994; Mayer, 1990).

[6]See also Eichengreen et al. (1996) and Frankel and Rose (1996).

[7]See Kaminsky (1999) and Kaminsky and Reinhart (1999) among others.

vance of this kind of empirical analysis. As shown above, in figure 9.12, an increasing quote of US listed firms displays an unbalanced financial structure. But this evidence is generally not so dissimilar from the rest of the industrialized world[8]. Even though a comprehensive analysis goes beyond the purposes of this work, some evidence regarding Italian corporate debt is reported below.

Figure 10.11 displays the trend of business debt as quote of GDP in Italy. The adopted proxy is the global credit accorded to firms. In only 8 years, the ratio recorded an increment of about 64%, a variation that can only be partially explained by a low level of interest rates. The second graph (figure 10.12) confronts the Gross Value Added and bad debt in Italy. Gross Value Added (GVA) is defined by Eurostat as output value at basic prices less intermediate consumption valued at purchasers' prices, calculated before consumption of fixed capital. The series has been cleaned of seasonal effects by means of moving average (on annual base), and de-trended by means of Hodrick-Prescott filter. Bad debt (in Italian "Sofferenze") is, for a bank or a financial institution, the credit that is unlikely to be paid back. Due to continuous changes in the definition of bad debt position, the graph covers only a brief period but, nonetheless all the limits of the adopted proxy[9], depicts a quite evident negative correlation with aggregate performances of the productive sector.

---

[8]See Steve Keen's www.debtdeflation.com/blogs.

[9]First of all the dimensional limit: only credits bigger than (or with securities bigger than) 75000 Euros are included.

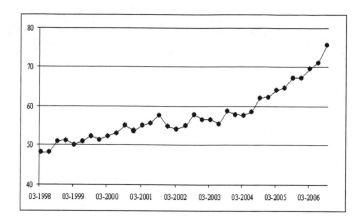

Figure 10.11 – Italy 1998-2006. Business debt (defined as "Accordato operativo") on GDP (per thousand). Source: elaboration on Bank of Italy and Eurostat Data.

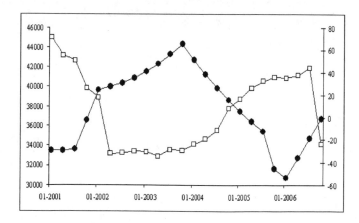

Figure 10.12 – Italy, 2001-2006. Bad debt (empty squares) and difference of gross added value from its trend (full circles, right axis). Source: elaboration on Bank of Italy and Eurostat data.

Chapter 11

Financial variables and firms' profitability: a hazard function analysis

In this chapter a hazard function analysis is performed on a set of European firms in order to identify a stochastic relationship among financial structure and profits. The relative proportions of debt and equity financing appear to influence expected profitability with a different degree for each nation. Within each country, relevant differences are recorded among listed and non listed firms. These results highlight the role of institutional factors, in particular related to credit and stock markets, in reducing informational asymmetries between investors and managers. The cross-sectional study is performed by means of degradation analysis, an engineering tool new in economics.

The structure of the chapter is the following: section 11.2 introduces the methodology used in the analysis; section 11.3 describes the particular model adopted to apply the methodological framework; section 11.4 shows the results of the analysis and section 11.5 closes.

11.1 Introduction

Choices of firms as regards sources of financing and their effects have been widely explored by theoretical and empirical literature. Given that conditions of Modigliani-Miller theorem (1958) do not hold in real world markets, the identification of a proper theoretical framework, the evaluation of firms' behaviors in terms of economic performances and the empirical tests of the proposed models have been matters of question for fifty years. Two main approaches can be recognized. First, at least chronologically, the *trade-off* theory (Kraus and Litzenberger, 1973) postulates that firms target an optimal financial equilibrium by equating marginal benefits of taxation of debt and marginal costs of bankruptcy. The second is the *pecking order* theory (Myers, 1984) that relies upon the already cited financing hierarchy hypothesis proposed by Myers and Majluf (1984). According to the theory, firms would at first use retained profits and internal funds to finance investments and, then, subordinately, prefer new debt instead of emitting equities. The two theories are founded on two opposite backgrounds: a perfect rationality approach for the first and asymmetric information for the other.

The static trade off (Bradley et al., 1984) defines the firms' object as a single-period balance between tax benefits and bankruptcy costs of debt. Arguments of firms' object function are the level of financial distress, the structure of corporate and personal taxation and the general uncertainty. In a dynamic context this strategy gives rise to gradual adjustments to the desired level of leverage. In this perspective a relevant role is played by transactional costs (Fischer et al., 1989), that prevent a firm from immediately adjusting its financial structure.

The suggested explanations of the pecking order theory address the informational problems in capital markets and, in particular, the agency problem that arises between external stockholders and firm's management. Jensen and Meckling (1976) identify the agency costs that originate in recurring to external finance. Stockholders and debtholders may demand covenants to prevent management from undertaking hazardous projects or, in general, from jeopardizing their interests. These limitations of manager's initiatives plus the direct cost of monitoring constitute the agency costs. The model of Myers and Majluf (1984) addresses the informational asymmetries between investors and management. They model financial markets along the lines of the market of "lemons" introduced by Akerlof (1970). Investors cannot distinguish "good" and "bad" projects and they apply the same risk premium to all. As a consequence, better projects, if financed with external funds, are undervalued. The straightforward conclusion is that, in financing a secure investment, managers first use internal sources. Among external sources debt should be preferred. Given the stricter controls to which a project is subject by potential creditors, the emission of equities can be interpreted by (actual and potential) stockholders as an attempt to hide hazardous managers' behaviour, especially if it determines a reduction in the value of the existing stocks. The concession of debt is usually subordinated to an accurate evaluation of the financed ventures and of balance sheets, that would prevent managers from undertaking hazardous projects.

A number of papers empirically tested the two approaches. According to the exhaustive review on the topic by Frank and Goyal (2008), that list the basic stylized facts, a conclusive proof is far from being reached. Some of the findings are contradictory among each other and for most of them it is not possible to formulate an unambiguous interpretation that supports a theoretical explanation rather than the other. Despite wide evidence in favor of the pecking order approach, the debate is still very lively.

Fama and French (2002), investigating the relationship among leverage with dividends and profits, found that, for these issues, there is no actual divergence between the two theories. Indeed, empirical results can be convincingly interpreted by both of them. Shyam-Sunder and Myers (1994), comparing the theories, found support for pecking order in analyzing US firms capital structure. Their work has been criticized by Chirinko and Singha (2000), according to whom neither the pecking order nor static trade off models can explain the evidence. One of the best known empirical verifications of the pecking order models is the one by Fazzari et al. (1988). They found significant confirmation of theory's prediction about preference for internal funds, detecting a positive correlation among cash flow and investment and lower payouts for financially constrained firms. With regards to strategic assets, Kochar (1997) found specific evidence about better performances for firms that prefer internal resources. As regards the sources of the hierarchy, two works of Oliner and Rudebusch (1989; 1992) emphasized the role of informational asymmetries, instead of explanations based on transactional costs, that were prominent in earlier works[1]. The first demonstrates that transactional costs of debt and equities account for a relevant percentage of small emission of debts and equities while this percentage becomes negligible in case of big emissions. In the second, they discovered that the known correlation among cash flow and investment appears to be closer for firms supposed to face relatively severe information asymmetries. However, Myers (2003) shows that also the agency cost theory can be included in a pecking order framework.

With specific reference to profits, there is evidence of a negative relation between leverage ratio and profitability of firms (Fama and French, 2002). This fact appears to be in line with the pecking order theory while implications are ambiguous for the trade off. The dynamic trade off predicts a higher debt ratio for profitable firms, due to the reduced costs of bankruptcy (Jensen, 1986) while, considering profitability as a proxy for growth, static theory implications are compatible with such evidence (Frank and Goyal, 2008). Chen and Zhao (2005) checked the suitability of trade-off with costly adjustment explanations for firms' issuances decisions. Their work confirms the evidence of higher profits for underlevered firms, but does not obtain

---

[1]See also Samuel (1996) for a comparison with managerial theories

any evidence in favor of trade-off explanations, actually leaving the question open.

This chapter investigates the relationship among financial soundness and perspective profitability of firms, but with a brand new approach. The intended contribution to the existing debate is not mainly to reconcile reality with one of the existing theories. As shown, the topic has been largely investigated using static and dynamic panel data analysis, as in the cited works of Fama and French (2002) and Chen and Zhao (2005). This methodology cannot get rid of heterogeneity of firms as concerns the relative different weights of debt that clearly influence the impact of a marginal variation in capital structure. The hazard function analysis takes into consideration this dynamic aspect. Moreover it allows a reformulation of the problem in a probabilistic framework that appears more suitable to heterogeneous and evolving firms.

For what regards the object, the analysis is performed separately for listed and non listed firms in different countries in order to highlight the impact of institutional factors on informational asymmetries. The starting hypothesis is that listed firms are expected to face less constraining informative asymmetries respect to non listed. Hazard function analysis, estimating a proxy for the size of these disparities for each country, permits to compare the roles played by institutional factors in determining the contexts in which firms decide about their capital structures. In this perspective, firms are considered as objects that can break, in order to make inference about the maximum level of debt they can stand maintaining a positive profit.

To summarize, I will try to answer three specific questions: first, if the probability for a firm to be profitable changes with leverage; second, how the presence of informational asymmetries between management and investors impacts the dependence of perspective performances upon leverage; third, if and how institutional factors influences the relationship.

11.2   Methodology

This part of the work introduces the theoretical framework adopted to answer the three questions in which the empirical analysis is articulated.

The first general question about the dependence of profit's probability upon leverage is dealt with by means of hazard function techniques. These inference instruments permit to estimate the rate of survival, referred to a specified probability function, over a certain point x once that point has been reached.

Formally, hazard function $h(x)$ is defined as the ratio of the probability function $p(x)$ to the survival function $S(x)$, and then it is given by:

$$h(x) = \frac{p(x)}{S(x)} = \frac{p(x)}{1 - F(x)} \tag{11.1}$$

where $F(x)$ is the distribution function. Hazard and survival analysis are definitely not new in economics. Some applications of hazard functions have been devoted for duration of unemployment (Lancaster, 1979) and business cycle fluctuations (Sichel, 1991). Other works, as the cited Di Guilmi et al. (2005) which studied relative magnitude of business cycles phases, adopted alternative non-negative explanatory variables instead of time[2].

As regards the second question, the analysis lies on the hypothesis that the public quotation of a firm reduces the informational asymmetries between investors and management. Public traded companies are subject to a stricter set of rules, e. g. for what regards publication of balance sheets, discipline of corporate boards, buying and selling of quotes. These rules define the informative set available for investors, which is wider in case of publicly traded companies.

Moreover, this legal discipline is different from country to country. For this reason, in order to take into exam the third question, investigation is performed on a set of seven countries, composing a homogeneous sample of companies for each of them. Therefore, the analysis considers, for each country, two dynamic panels, one for listed firms and another for non listed ones. Details about data and sampling are provided in section 11.3.

The main problem is how to deal with these micro data in order to obtain a usable input for the hazard function, saving, at the same time, the heterogeneity of firms. More in particular, the two-dimensional matrix of units and time has to be reduced to one dimension in order to get a single synthetic value for each unit to perform survival analysis. Regarding phenomenological aspects, the predictability of profit margin of a firm is analogous to the problem of testing products performed in failure analysis. Failure analysis measures the probability for a group of products to observe a certain number of "failures", i. e. damages or losses. More specifically, I make use here of *degradation analysis*, a tool widely adopted in engineering to test and measure the breaking point of a class of products. It basically consists in a cross-section analysis of survival times for homogeneous products (for example steel plates), put into a functional relationship with a measurable factor whose relevance has to be tested (for example the length of a crack in the plate). It is called degradation analysis just because the basic hypothesis is

---

[2]See Bhattacharjee (1988) for a survey.

that the subject of the experiment deteriorates as time passes. The degree of deterioration is considered as dependent on the pre-defined measurable variable. Formally, given a set of data

$$\{Y_j(t), \ j \in J, \ t \in T\}$$

where $Y_j(t)$ is the degradation data for unit j at time t (length of the crack in the j-th steel plate in our example), the condition for it to be suitable for degradation analysis is:

$$\mathbb{E}\left[Y(t)\right] < \mathbb{E}\left[Y(t+1)\right] \quad \forall t \in T \tag{11.2}$$

That is to say that expected degradation values increase monotonically with time (Chao, 1999). Fixing a critical threshold of the observable (the critical length of the crack that makes the plate break), it is possible to perform an interpolation on the series of observations T: $Y_j(t)$, $Y_j(t+1)$, ..., $Y_j(T)$, to estimate for each unit j the expected survival time. In this way for each micro unit it is possible to obtain a single reference value, making the two dimensional matrix collapse into a vector. The vector of survival times constitutes the input of the hazard function.

Thus, it is possible to change the reference variable from time, which is not relevant for this work's purpose, to another variable, provided that the condition (11.2) holds, in order to obtain the vector of critical values to perform the hazard function analysis.

11.3   The model

The data analyzed here are taken from the commercial data set Amadeus and organized in balanced panels, one for each considered country (namely: France, Germany, Greece, Italy, Spain, Sweden, United Kingdom), with different time spans, over the period 1992-2003, chosen on the base of the availability of a sufficient number of continuing firms. The panel is built to have the longest possible time span, in order to use enough points to perform degradation regressions, and the highest possible number of firms for each subset, maintaining at the same time a comparable composition by industry among countries. Since degradation regressions are performed on each unit, only continuing firms are included in the sample in order to get a sufficient number of observations in all regressions. Only joint-stock companies are taken into exam in order to compare firms that have access to both forms of financing (credit market and stocks), reducing the heterogeneity that may originate from the different availability of financing options. Firms are then divided between listed and non listed. The adoption of public quotation as grouping criteria aims to evaluate the impact of informational asymmetries in determining the dependence of corporates' performances with reference to their capital structure.

Equity ratio is taken as independent variable for degradation regressions. Return on assets, calculated as profit before taxes on total assets, is chosen as proxy for firms' profitability. Profit before taxes is calculated as operating plus financial results and this definition is homogeneous for all national accounting systems in the examined countries[3]. The inclusion of taxes might introduce cross-biases due to different accounting definitions and distortions in the regressions due to changes of fiscal rules in the observation period. The relationship between equity ratio and return on assets respects the condition (11.2), as displayed in figure 10.10 for Italy. Analogue results have been obtained for the other examined countries.

The following step consists in determining the critical value of the dependent variable, i. e. the failure level of return of assets. It is fixed at 0, considering the failure as a loss or negative profit. It is then possible to run degradation regression in order to estimate, for each firm included in the sample, a synthetic indicator to perform, subsequently, a hazard function based investigation. This first stage of investigation allows to estimate the critical level of equity ratio at which return on assets is expected to be null. The adopted regression model is:

$$y_{it} = \gamma_i x_{it} + \alpha_i + \epsilon_{it} \tag{11.3}$$

where x_{it} is the equity ratio for firm i at time t and y_{it} is the return on assets for each observation. γ_i and α_i are the estimated parameters while ϵ_{it} is the residual. The procedure is repeated for each firm. Then the values x_i for which $y_i = 0$ are grouped in vectors containing the critical equity ratio's levels for listed and non listed firms in each country.

A hazard function analysis is then performed on these populations of estimated values. The most used model in economic literature is the Weibull parametric hazard function (Lancaster, 1979). One of the advantages of the Weibull hazard model is that it permits to infer, by a synthetic indicator, the sign of the relationship among variables. Preliminarily, the suitability of the model to the examined data sets is checked. Weibull distribution returns a very good fit for all the populations of critical values, as derived by the degradation analysis. According to Lancaster (1979), the model is specified in the following way:

$$h_w(y) = \sigma\beta y^{\beta-1} \tag{11.4}$$

---

[3]Only for Germany it is defined as the net operating profit and, therefore, for the elaboration, it has been summed with the financial result.

being β the shape parameter of the theoretical Weibull density and $\sigma = \eta^{-\beta}$ a function of the scale parameter η. This function returns the conditional probability of obtaining a value $y > y_0$ once a value y_0 has been reached. This probability is an increasing (decreasing) function of y if β is bigger (smaller) than 1. I did not adopt here the three-parameters Weibull model (Mudholkar et al., 1996) since in all cases, the estimations of the location parameter of the Weibull distribution return a value not significantly different from 0. The estimations are performed by means of y rank regression (Abernethy, 1996).

11.4   Empirical results

By the analysis of the estimates of parameter β, that are reported in table 11.1, it is possible to obtain some insights on the stochastic relationship under exam. In particular, for $\beta > 1$ the effect of financial soundness in determining firm's profitability appears as progressively more relevant, while the occurrence of $\beta < 1$ means that the probability of observing a positive profit is negatively related with equity ratio. A β equal to 1 reveals statistical independence.

For all countries, except United Kingdom, β is bigger for non listed firms (i.e. return on assets is stricter dependent on equity ratio). For publicly traded companies of five on seven countries an increase in the level of debt raises the probability of obtaining positive profits. The exceptions are Italy (figure 11.1), whose βs are both larger than 1, and the United Kingdom, which shows statistical independence. Non listed firms, in five cases out of seven, show positive dependence. Exceptions are Spain and Sweden, whose βs are lower than 1 but noticeably higher with respect to estimations for the non listed. For France, Germany and Greece hazard function displays an increasing failure rate for non listed firms and a decreasing failure rate for listed ones (see figure 11.2 for France). Italy and Spain exhibit divergent behaviors: in the latter case for both classes of companies hazard rate of return on assets appears to be positively dependent on the level of debt (also for non listed β is significantly lower than 1), while, as regards Italy, even for publicly traded firms, equity ratio has an increasing positive effect on profit margins.

With reference to the theory, at first sight, listed firms appear to confirm trade off implications while the pecking order theory can explain results for the non listed. However, from a different perspective, a pecking order interpretation may emerge as more appropriate and complete for both groups. Indeed, according to the pecking order hypothesis, in presence of informational asymmetries, debt should be preferred among external financing op-

State	Listed firms	Non listed firms
France	0.9752 ± 0.0221	1.0409 ± 0.0234
Germany	0.9168 ± 0.0186	1.2236 ± 0.0963
Greece	0.9053 ±0.0854	1.0574 ± 0.0304
Italy	1.1435 ± 0.0822	1.1618 ± 0.0163
Spain	0.9164 ± 0.1355	0.9806 ± 0.0249
Sweden	0.9235 ± 0.0696	0.9812 ± 0.0108
United Kingdom	1.0003 ± 0.0562	1.0712 ± 0.0227

TABLE 11.1 – *Empirical estimates of β.*

tions. Due to legal constraints (e. g. stricter control of balance sheets), the public quotation of a firm reduces the asymmetry of information among managers and investors. As a consequence publicly traded companies can obtain credit with more facility and with lower costs than private ones. As reported in Smith and Warner (1979), the service on debt may be higher for firms that face larger informational asymmetries or limitations in varying capital structures, making their profitability more dependent on their financial structure. Both classes of firms may prefer debt rather than equity and both face informational barriers. But when informational asymmetries are larger, the increasing costs of debt augments the probability of a negative profit.

From this point of view, it is also possible to cast some light on the wide differences recorded between countries. Credit markets and access to quotation are differently disciplined in each country and, subsequently, these informative mechanisms works with different efficacy. Extreme examples are Germany and Italy. For Germany, the different situation faced by the two types of firms is testified by the huge gap between the estimations. For Italian firms, public quotation does not imply a higher level of trust for banks and investors and therefore a lower premium on risk: listed companies are not less dependent than others on their level of debt.

Other institutional factors can be addressed to explain the dissimilarities between and within countries, but their effective role in determining the outcomes is difficult to weigh. For most of them, unambiguous evidence cannot be found in earlier literature.

An order of possible motivations refers to the discipline of firms' exits. It determines the costs of bankruptcy and, in this way, the cost of debt[4]. In this

---

[4]The discipline for mergers and acquisitions is connected to the explanations based on bankruptcy costs. According to Maksimovic and Phillips (1998), they permit to avoid at least the direct costs of bankruptcy, reducing the marginal cost of debt.

perspective, it appears of particular interest that one of the countries with the most flexible bankruptcy law, the United Kingdom, reveals statistical independence of level of debt and profit for non listed companies and no significant differences among listed and non listed ones.

An important factor of heterogeneity in firms' financing options investigated by the literature is the firms' size. Different works on the same topic found contradictory results. For example, the sign of the relationship between size and equity ratio is positive according to Kester (1986) and negative for Rajan and Zingales (1995). As displayed in table 11.3 the average size of firms differ remarkably among countries, but the indications regarding the estimations of βs are ambiguous. The disparities among countries do not seem to be related to the average size of sampled firms.

Part of the differences in the estimations can be addressed also to the different disciplines of managerial compensation (Dybvig and Zender, 1991) that could reduce adverse selection of risky projects. In this perspective, a reduction in informational barriers could reduce the cost of financing but it is not clear in which direction.

Another possible explanation lies on Fischer's separation theorem, according to which, in the presence of incomplete markets, the ownership of the debt assumes relevance in firms' decisions. In a globalized market, where the share of transnational hedge funds in financial investments is sharply rising, the relevance of such an argument can be reasonably considered as vanishing.

To conclude, this non exhaustive listing highlights the importance of informational imperfections in financial markets as the main determinant of the detected stochastic relationships. The results of comparative analysis of the two types of firms verify the stricter dependence of non listed firms on their own capital. These outcomes do not appear to be biased by the presence, in the same cluster of analysis, of firms of different dimensions.

11.5   Concluding remarks

This chapter analyzes, by means of degradation analysis and a two parameter Weibull hazard model, the dynamics of return on assets conditioned on equity ratio, of listed and non listed firms for a group of seven European countries. The results display significant differences among countries, even if, for all of them, the dependence on equity ratio is never smaller for non listed companies. These evidence may be explained by the peculiarities, at behavioral and institutional levels, of each nation as regards credit market and access to quotation, and by the capacity of these regulations to reduce asymmetric information (and, in this way, the premium on credit risk) for investors.

The analysis allows to answer the three questions that motivate it. First, the probability of firms to be profitable is linked to their financial situation. Second, information asymmetries play a role in determining the perspective profitability of units through costs of financing. Finally, the study shows that this role appears to be qualitatively and quantitatively different for each nation, pointing out the impact of institutional factors.

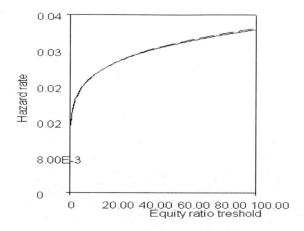

Figure 11.1 – Hazard function for return on assets in Italy (dashed line refers to non quoted firms).

Country	Non listed	Listed
France	13551	697
Germany	1238	278
Greece	3142	201
Italy	5307	1665
Spain	4219	86
Sweden	6503	192
UK	681	385

TABLE 11.2 – *Number of firms included in the analysis.*

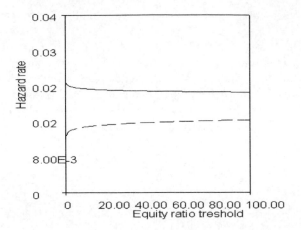

Figure 11.2 – Hazard function for return on assets in France (dashed line refers to non quoted firms).

Country	Non listed	Listed	Mean
France	161.11	278.14	166.95
Germany	2413.57	12741.05	4886.41
Greece	142.24	536.43	204.14
Italy	78.74	131.34	80.07
Spain	68.13	67.08	68.08
Sweden	226.69	5882.86	415.87
United Kingdom	1829.87	7251.35	3948.31

TABLE 11.3 – *Average number of employees in the time span of observation. Source: elaboration on Amadeus data.*

Bibliography

ABERNETHY, R. B. (1996): *The New Weibull Handbook*, Gulf Publishing Company, second edn.

ADEMA, Y. and STERKEN, E. (2005): Monetary Policy Rules: From Fisher to Svensson, Taylor, and Woodford, *Financial and Monetary Studies*, **23**(4).

AKERLOF, G. A. (1970): The Market for 'Lemons': Quality Uncertainty and the Market Mechanism, *The Quarterly Journal of Economics*, **84**(3): pp. 488–500.

ALTMAN, E. I. (1968): Financial Ratios, Discriminant Analysis and the Prediction of Corporate Bankruptcy, *Journal of Finance*, **23**(4): pp. 589–609.

ANDERSON, P. W., ARROW, K. J. and PINES, D. (eds.) (1988): *The Economy as an Evolving Complex System*, Sante Fe Institute.

AOKI, M. (1996): *New approaches to macroeconomic modeling*, Cambridge University Press.

AOKI, M. (2002): *Modeling aggregate behaviour and fluctuations in economics*, Cambridge University Press.

AOKI, M. and YOSHIKAWA, H. (2006): *Reconstructing Macroeconomics*, Cambridge University Press.

ARTHUR, W. B. (1999): Complexity and the Economy, *Science*, **284**(5411): pp. 107–109, doi:10.1126/science.284.5411.107.

ARTHUR, W. B., ERMOLIEV, Y. and KANIOVSKI, Y. M. (1987): Path Dependent Processes and the Emergence of Macro-Structure, *European Journal of Operational Research*, **30**: pp. 294–303.

AUSLOOS, M., MISKIEWICZ, J. and SANGLIER, M. (2004): The Durations of Recession and Prosperity: Does Their Distribution Follow a Power or an Exponential Law?, http://arxiv.org/archive/cond-mat/0403143.

AXELROD, R. (1997): Advancing the Art of Simulation in the Social Sciences, Working Papers 97-05-048, Santa Fe Institute.

AXTELL, R. (2001): Zipf distribution of U.S. firm sizes, *Science*, **293**(5536): pp. 1818–1820.

AXTELL, R., AXELROD, R., EPSTEIN, J. M. and COHEN, M. D. (1996): Aligning simulation models: A case study and results, *Computational & Mathematical Organization Theory*, **1**(2): pp. 123–141.

AZLAROV, T. and VOLODIN, N. (1986): *Characterization Problems Associated with the Exponential Distribution*, Springer Verlag.

BAK, P. (1997): *How Nature Works*, Springer-Verlag, New York.

BALIAN, R. (1991): *From Microphysics to Macrophysics, Volume I*, Berlin/Heidelberg/New York: Springer-Verlag.

BEAN, C. R. (2004): Asset Prices, Financial Instability, and Monetary Policy, *American Economic Review*, **94**(2): pp. 14–18.

BEAVER, W. (1967): Financial Ratio as Predictors of Failure, Empirical Research in Accounting: Selected Studies 1966, *Journal of Accounting Research*, **S4**: pp. 71–111.

BELLMAN, R. E. (1961): *Adaptive control processes: A guided tour*, Princeton University Press, Princeton, New Jersey.

BERNANKE, B. and GERTLER, M. (1989): Agency Costs, Net Worth, and Business Fluctuations, *American Economic Review*, **79**(1): pp. 14–31.

BHATTACHARJEE, M. C. (1988): Reliability Ideas and Applications in Economics and Social Sciences, in: KRISHNAIAH, P. R. and RAO, C. R. (eds.), *Handbook of Statistics*, vol. 7, chap. 11, pp. 175–213, Elsevier Science Publishers B. V.

BRADLEY, M., JARRELL, G. A. and KIM, E. H. (1984): On the Existence of an Optimal Capital Structure: Theory and Evidence, *Journal of Finance*, **39**(3): pp. 857–78.

BROOK, D. (1964): On the distinction between the conditional probability and the joint probability approaches in the specification of nearest-neighbour systems., *Biometrica*, **51**: pp. 481–483.

BURNS, A. and MITCHELL, W. (1946): *Measuring Business Cycles*, National Bureau of Economic Research, New York.

CABALLERO, R. J. and ENGEL, E. M. (1991): Dynamic (S, s) Economies, *Econometrica*, **59**(6): pp. 1659–86.

CHAO, M.-T. (1999): Degradation Analysis and Related Topics: Some Thoughts and a Review, *Proc Natl Sci Counc*, **23**(5): pp. 555–566.

CHEN, L. and ZHAO, X. (2005): Firm financing decisions, working paper, Michigan State University and Kent State University.

CHIRINKO, R. S. and SINGHA, A. R. (2000): Testing static tradeoff against pecking order models of capital structure: a critical comment, *Journal of Financial Economics*, **58**(3): pp. 417–425.

CLEMENTI, F., DI MATTEO, T. and GALLEGATI, M. (2006): The power-law tail exponent of income distributions, *Physica A: Statistical and Theoretical Physics*, (370).

CLIFFORD, P. (1990): Markov random fields in statistics, in: GRIMMETT, G. R. and WELSH, D. J. A. (eds.), *Disorder in Physical Systems. A Volume in Honour of John M. Hammersley*, pp. 19–32, Oxford: Clarendon Press.

COOLEY, T. and QUADRINI, V. (2001): Financial Markets and Firm Dynamics, *The American Economic Review*, **91**(5): pp. 1286–1310.

COX, D. and MILLER, H. (1996): *The Theory of Stochastic Processes*, Chapman and Hall.

DAVID, P. A. and FORAY, D. (1993): Percolation structures, Markov random fields and the economics of EDI standard diffusion, in: *Global telecommunications strategies and technological changes*, North-Holland.

DE LA LAMA, M. S., SZENDRO, I. G., IGLESIAS, J. R. and WIO, H. S. (2006): Van Kampen's expansion approach in an opinion formation model, *The European Physical Journal B*, **51**(3): pp. 435–442.

DELLI GATTI, D. (1999): Firms'Size and Monetary Policy: Some New Keynesian Reflections, in: GALLEGATI, M. and KIRMAN, A. (eds.), *Beyond the Representative Agent*, pp. 141–161, Elgar.

DELLI GATTI, D., DI GUILMI, C., GAFFEO, E. and GALLEGATI, M. (2004): Bankruptcy as an exit mechanism for systems with a variable number of components, *Physica A: Statistical Mechanics and its Applications*, **344**(1-2): pp. 8–13.

DELLI GATTI, D., DI GUILMI, C., GAFFEO, E., GIULIONI, G., GALLEGATI, M. and PALESTRINI, A. (2005): A new approach to business fluctuations: heterogeneous interacting agents, scaling laws and financial fragility, *Journal of Economic Behavior and Organization*, **56**(4): pp. 489–512.

DELLI GATTI, D., DI GUILMI, C., GALLEGATI, M. and GIULIONI, G. (2007): Financial Fragility, Industrial Dynamics, And Business Fluctuations In An Agent-Based Model, *Macroeconomic Dynamics*, **11**(S1): pp. 62–79.

DI GUILMI, C., GAFFEO, E. and GALLEGATI, M. (2004): Empirical results on the size distribution of business cycle phases, *Physica A: Statistical Mechanics and its Applications*, **333**(15 Feb): pp. 325–334.

DI GUILMI, C., GAFFEO, E., GALLEGATI, M. and PALESTRINI, A. (2005): International Evidence on Business Cycle Magnitude Dependence: An Analyisis of 16 Industrialized Countries, 1881-2000, *International Journal of Applied Econometrics and Quantitative Studies, Euro-American Association of Economic Development*, **2**(1): pp. 5–16.

DIEBOLD, F. X. and RUDEBUSCH, G. (1990): A Nonparametric Investigation of Duration Dependence in the American Business Cycle, *Journal of Political Economy*, **98**(3): pp. 598–616.

DOW, C. (1998): *Major recessions. Britain and the World, 1920-1995*, Oxford University press.

DUNNE, P. and HUGHES, A. (1994): Age, Size, Growth and Survival: UK Companies in the 1980s, *Journal of Industrial Economics*, **42**(2): pp. 115–40.

DYBVIG, P. and ZENDER, J. (1991): Capital structure and dividend irrelevance with asymmetric information, *Review of Financial Studies*, **4**(1): pp. 201–219(19).

EICHENGREEN, B., ROSE, A. and WYPLOSZ, C. (1996): Contagious Currency Crises: First Tests, *Scandinavian Journal of Economics*, **98**(4): pp. 463–84.

FAMA, E. F. and FRENCH, K. R. (2002): Testing Trade-Off and Pecking Order Predictions About Dividends and Debt, *Rev Financ Stud*, **15**(1): pp. 1–33.

FASANO, G. and FRANCESCHINI, A. (1987): A multi-dimensional version of the Kolmorogov-Smirnov Test, *Monthly Notices of the Royal Astronomical Society*, **225**: pp. 155–170.

FAZZARI, S. M., HUBBARD, G. and PETERSEN, B. C. (1988): Financing Costraints and Corporate Investments, *Brookings Papers on Economic Activity*, **1**: pp. 141–195.

FISCHER, E. O., HEINKEL, R. and ZECHNER, J. (1989): Dynamic Capital Structure Choice: Theory and Tests, *Journal of Finance*, **44**(1): pp. 19–40.

FISCHER, I. (1932): *Booms and Depressions*, New York, Adelphi.

FOLLMER, H. (1974): Random economies with many interacting agents, *Journal of Mathematical Economics*, **1**(1): pp. 51–62.

FORNI, M. and LIPPI, M. (1997): *Aggregation and the Microfoundations of Dynamic Macroeconomics*, Oxford University.

FOSTER, J. (2004): Why is Economics not a Complex Systems Science?, Discussion Papers Series 336, School of Economics, University of Queensland, Australia, available at http://ideas.repec.org/p/qld/uq2004/336.html.

FRANK, M. Z. and GOYAL, V. K. (2008): Trade-off and Pecking Order Theories of Debt, in: ECKBO, E. (ed.), *Handbook of Corporate Finance: Empirical Corporate Finance, Handbooks in Finance Series*, vol. 2, Elsevier/North-Holland.

FRANKEL, J. A. and ROSE, A. K. (1996): Currency Crashes in Emerging Markets: Empirical Indicators, CEPR Discussion Papers 1349, C.E.P.R. Discussion Papers.

FUJIWARA, Y. (2004): Zipf law in firms bankruptcy, *Physica A: Statistical and Theoretical Physics*, **337**(1-2): pp. 219–230.

FUJIWARA, Y., DI GUILMI, C., GALLEGATI, M., HAOYAMA, H. and SOUMA, W. (2004): Do Pareto-Zipf and Gibrat Law Hold True? An Analysis with Eruopean Firms, *Physica A: Statistical Mechanics and its Applications*, **335**(1-2): pp. pp. 197–216.

FULMER, J. G., MOON, J. E., GAVIN, T. A. and ERWIN, M. J. (1984): A Bankruptcy Classification Model For Small Firms, *Journal of Commercial Bank Lending*.

GABAIX, X. (2005): The Granular Origins of Aggregate Fluctuations, Meeting Papers 470, Society for Economic Dynamics.

GAFFEO, E., GALLEGATI, M. and PALESTRINI, A. (2003): On the size distribution of firms. Additional evidence from the G7 countries, *Physica A*, (1-2).

GALLEGATI, M. (2002): Una generalizzazione dell'approccio Greenwald-Stiglitz con imprese eterogenee, *Economia Politica*, (1).

GALLEGATI, M., PALESTRINI, A., DELLI GATTI, D. and SCALAS, E. (2006): Aggregation of heterogeneous interacting agent: the variant representative agent framework, *Journal of Economic Interaction and Coordination*, (1): pp. 5–19.

GERTLER, M. (1994): Financial Condition and Macroeconomic Behaviour, Tech. Rep. 10-14, NBER Reporter Summer.

GOLDFAJN, I. and VALDES, R. O. (1998): Are currency crises predictable?, *European Economic Review*, **42**(3-5): pp. 873–885.

GORMAN, W. M. (1953): Community Preference Fields, *Econometrica*, **21**(1): pp. 63–80.

GREENWALD, B. and STIGLITZ, J. E. (1990): Macroeconomic models with equity and credit rationing., in: HUBBARD, R. (ed.), *Information, Capital Markets and Investment.*, Chicago University Press, Chicago.

GREENWALD, B. and STIGLITZ, J. E. (1993): Financial markets imperfections and business cycles, *Quarterly journal of Economics*, **108**(1): pp. 77–114.

GREENWALD, B. C., STIGLITZ, J. E. and WEISS, A. (1984): Informational Imperfections in the Capital Market and Macro-Economic Fluctuations, *American Economic Review*, **LXXIV**: pp. 194–99.

GRENDAR, M. and GRENDAR, M. (2002): Why maximum entropy? A non-axiomatic approach, in: FRY, R. L. (ed.), *Bayesian Inference and Maximum Entropy Methods in Science and Engineering, American Institute of Physics Conference Series*, vol. 617, pp. 375–378.

GRIMMETT, G. and STIRZAKER, D. (1992): *Probability and Random Processes*, Clarendon Press, Oxford, second edn.

HAMMERSLEY, J. M. and CLIFFORD, P. (1971): *Markov field on finite graphs and lattices*, unpublished.

HARDING, D. and PAGAN, A. (2002): Dissecting the business cycle: a methodological investigation, *Journal of Monetary Economics*, **49**(2): pp. 365–381.

HENNESSY, C. A. and WHITED, T. M. (2007): How Costly Is External Financing? Evidence from a Structural Estimation, *Journal of Finance*, **62**(4): pp. 1705–1745.

HILL, B. M. (1975): A simple general approach to inference about the tail of a distribution., *The Annals of Statistics*, **3**(5): pp. 1163–1174.

HINICH, M. J., FOSTER, J. and WILD, P. (2006): Structural change in macroeconomic time series: A complex systems perspective, *Journal of Macroeconomics*, **28**(1): pp. 136–150.

HODRICK, R. J. and PRESCOTT, E. C. (1998): Postwar US business cycles: an empirical investigation, *Journal of Money, Credit and Banking*, **29**(1).

IJIRI, Y. and SIMON, H. (1977): *Skew Distributions and the Sizes of Business Firms*, North-Holland, New York.

INTERNATIONAL MONETARY FUND, I. M. F. (2002): *World Economic Outlook*, IMF, Washington.

JAYNES, E. T. (1957): Information Theory and Statistical Mechanics, *Phys Rev*, **106**(4): pp. 620–630, doi:10.1103/PhysRev.106.620.

JAYNES, E. T. (1968): Prior probabilities., *IEEE Trans Syst Sci Cybern*, **SSC-4**(3).

JAYNES, E. T. (1979): Concentration of distributions at entropy maximum, in: R.D., R. (ed.), *Jaynes E.T.: papers on probability statistics and statistical physics*, pp. 1–20, D. Reidel, Dordrecht.

JENSEN, M. (1986): Agency costs of free cash flow, corporate finance, and takeovers, *American Economic Review*, **76**(2): pp. 323–329.

JENSEN, M. C. and MECKLING, W. H. (1976): Theory of the Firm: Managerial Behavior, Agency Costs and Ownership Structure, *Journal of Financial Economics*, **3**(4): pp. 305–360.

JORGENSON, D. W., LAU, L. J. and STOKER, T. M. (1982): The trascendental Logarithmic Model Of Aggregate Consumer Behavior, in: BASNAM, R. and RHODES, G. (eds.), *Advances in Econometrics*, Greewich, Conn, JAI Press.

KAMINSKY, G. L. (1999): Currency and Banking Crises - The Early Warnings of Distress, IMF Working Papers 99/178, International Monetary Fund, available at http://ideas.repec.org/p/imf/imfwpa/99-178.html.

KAMINSKY, G. L. and REINHART, C. M. (1999): The Twin Crises: The Causes of Banking and Balance-of-Payments Problems, *American Economic Review*, **89**(3): pp. 473–500.

KEEN, S. (2001): *Debunking Economics: The Naked Emperor of the Social Sciences*, Sydney: Pluto Press.

KESTER, W. C. (1986): Capital and Ownership Structure: A Comparison of United States and Japanese Manufacturing Corporations, *Financial Management*, **15**(1): pp. 5–16.

KEYNES, J. M. (1936): *The General Theory of Employment, Interest and Money*, London: Macmillan.

KINDLEBERGER, C. (2005): *Manias, Panics, and Crashes: A History of Financial Crises*, Wiley, fifth edn.

KIRMAN, A. P. (1992): Whom or What Does the Representative Individual Represent?, *Journal of Economic Perspectives*, **6**(2): pp. 117–36.

KIYOTAKI, N. and MOORE, J. (1997): Credit Cycles, *Journal of Political Economy*, **105**(2): pp. 211–48.

KOCHAR, R. (1997): Strategic Assets, Capital Structure, and Firm Performance, *Journal of Financial And Strategic Decisions*, **10**(3): pp. 23–36.

KRAUS, A. and LITZENBERGER, R. H. (1973): A State-Preference Model of Optimal Financial Leverage, *Journal of Finance*, **28**(4): pp. 911–22.

KREGEL, J. (1998): Yes, "It" Did Happen Again - A Minsky Crisis Happened in Asia, Macroeconomics 9805017, EconWPA, available at http://ideas.repec.org/p/wpa/wuwpma/9805017.html.

KUBO, R., TODA, M. and HASHITUME, N. (1978): *Statistical Physics II. Non Equilibrium Statistical Mechanics*, Springer Verlag Berlin.

LANCASTER, T. (1979): Econometric Methods for the Duration of Unemployment, *Econometrica*, **47**: pp. 939–56.

LANDINI, S. (2005): *Modellizzazione stocastica di grandezze economiche con un approccio econofisico*, Ph.D. thesis, University Bicocca, Milan.

LANDINI, S. and UBERTI, M. (2008): A statistical mechanic view of macrodynamics in economics, *Computational Economics*, forthcoming.

LEWBEL, A. (1992): Aggregation with Log-Linear Models, *Review of Economic Studies*, **59**(3): pp. 635–42.

LINDAHL, E. (1939): *Studies in the Theory of Money and Capital*, George Allen & Unwin Ltd.

LINTNER, J. (1971): Corporate Finance: Risk and Investment, in: FERBER, R. (ed.), *Determinants of Investment Behavior*, New York: NBER.

LIOSSATOS, P. (2004): Statistical Entropy in General Equilibrium Theory, Working Papers 0414, Florida International University, Department of Economics.

LUCAS, R. E. J. (1977): Understanding Business Cycles, *Carnegie-Rochester Conference Series on Public Policy*, (5): pp. 7–29.

LUCAS, R. E. J. (1990): Why Doesn't Capital Flow from Rich to Poor Countries?, *American Economic Review*, **80**(2): pp. 92–96.

MAKSIMOVIC, V. and PHILLIPS, G. (1998): Asset Efficiency and Reallocation Decisions of Bankrupt Firms, *Journal of Finance*, **53**(5): pp. 1495–1532.

MANDELBROT, B. (1963): New Methods in Statistical Economics, *Journal of Political Economy*, **LXXI**(5).

MANKIW, N. G. (1985): Small Menu Costs and Large Business Cycles: A Macroeconomic Model, *The Quarterly Journal of Economics*, **100**(2): pp. 529–38.

MAYER, C. (1990): Financial systems, corporate finance and corporate finance, in: HUBBARD, R. G. (ed.), *Asymmetric information, corporate finance and investment*, Chicago: The University of Chicago Press.

MINSKY, H. (1963): Can "It" Happen Again?, in: CARSON, D. (ed.), *Banking and Monetary Studies*, Richard D Irwin, Homewood, reprinted in Minsky (1982).

MINSKY, H. (1982): *Inflation, recession and economic policy*, New York: ME Sharpe.

MODIGLIANI, F. and MILLER, M. (1958): The cost of capital, corporation finance and the theory of investment, *American Economic Review*, **XLVIII**(3): pp. 261–97.

MUDHOLKAR, G. S., SRIVASTAVA, D. K. and KOLLIA, G. D. (1996): A generalization of the Weibull distribution with application to the analysis of survival data, *Journal of the American Statistical Association*, **91**(436): pp. 1575–1583.

MYERS, S. C. (1984): Capital Structure Puzzle, NBER Working Papers 1393, National Bureau of Economic Research, Inc.

MYERS, S. C. (2003): Financing of corporations, in: CONSTANTINIDES, G., HARRIS, M. and STULZ, R. M. (eds.), *Handbook of the Economics of Finance*.

MYERS, S. C. and MAJLUF, N. S. (1984): Corporate Financing and Investment Decisions when Firms Have Information that Investors do not have, *Journal of Financial Economics*, **13**(2): pp. 187–221.

NEWMAN, M. E. J. (2005): Power laws, Pareto distributions and Zipfs law, *Contemporary Physics*, **46**(5).

OKUYAMA, K., TAKAYASU, M. and TAKAYASU, H. (1999): Zipfs Law in Income Distribution of Companies, *Physica A*, **269**(1): pp. 125–131.

OLINER, S. D. and RUDEBUSCH, G. D. (1989): Internal finance and investment: testing the role of asymmetric information and agency costs, Working Paper Series / Economic Activity Section 101, Board of Governors of the Federal Reserve System (U.S.), available at http://ideas.repec.org/p/fip/fedgwe/101.html.

OPPER, M. and SAAD, D. (2001): *Advanced Mean Field Methods: Theory and Practice*, The MIT Press. Cambridge, MA.

ORMEROD, P. and MOUNFIELD, C. (2001): Power law distribution of the duration and magnitude of recessions in capitalist economies: breakdown of scaling, *Physica A: Statistical Mechanics and its Applications*, **293**(573).

PALESTRINI, A. (2000): *Aggregazione tra Individui ed Interazione in Macroeconomia*, Ph.D. thesis, Universitá degli Studi di Ancona.

PESARAN, M. H. (2000): On the Aggregation of Linear Dynamic Models, mimeo.

POPPER, K. (2002): *All Life is Problem Solving*, Routledge.

RAJAN, R. G. and ZINGALES, L. (1995): What Do We Know about Capital Structure? Some Evidence from International Data, *Journal of Finance*, **50**(5): pp. 1421–60.

REINHART, C. M. and ROGOFF, K. S. (2004): Serial Default and the "Paradox" of Rich to Poor Capital Flows, Working Paper 10296, National Bureau of Economic Research.

RISKEN, H. (1989): *Fokker-Planck equation. Method of solutions and applications.*, Berlin: Springer Verlag.

ROCHA, L. M. (1999): *BITS: Computer and Communications News. Computing, Information, and Communications Division*, Los Alamos National Laboratory.

ROSS, S. (2003): *Introduction to probability models*, Academic Press.

ROSSER, J. B. (ed.) (2004): *Complexity in economics*, Edward Elgar Pub.

RUBENSTEIN, M. (1975): Security Market Efficency in an Arrow-Debreu Economy, *American Economic Review*, **65**(5): pp. 812–24.

SAMUEL (1996): Internal finance and investment: another look, *World Bank: Policy, Research working paper; WPS 1663*.

SETSER, B., ALLEN, M., KELLER, C., ROSENBERG, C. B. and ROUBINI, N. (2002): A Balance Sheet Approach to Financial Crisis, IMF Working Papers 02/210, International Monetary Fund, available at http://ideas.repec.org/p/imf/imfwpa/02-210.html.

SHYAM-SUNDER, L. and MYERS, S. C. (1994): Testing Static Trade-off Against Pecking Order Models of Capital Structure, Working Paper 4722, National Bureau of Economic Research.

SICHEL, D. E. (1991): Business cycle duration dependence: a parametric approach, *Review of Economics and Statistics*, (2): pp. 254–260.

SMALLWOOD, R. H. (1996): A two-dimensional Kolmorogov-Smirnov test for binned data, *Phys Med Biol*, **41**(1): pp. 125–135.

SMITH, C. and WARNER, J. (1979): On Financial Contracting: An Analysis of Bond Covenants, *Journal of Financial Economics*, **7**(2): pp. 117–161.

STIGLITZ, J. E. (1982): Information and Capital Markets, NBER Working Papers 0678, National Bureau of Economic Research, Inc.

STIGLITZ, J. E. (2000): What I learned at the World Economic Crisis, *The New Republic*, (17 Apr): pp. 56–60.

STIGLITZ, J. E. and WEISS, A. (1981): Credit Rationing in Markets with Imperfect Information, *American Economic Review*, **71**(3): pp. 393–410.

STOKER, T. M. (1984): Completeness, Distribution Restriction, and the Form of Aggregate Functions, *Econometrica*, **52**(4): pp. 887–907.

STOKER, T. M. (1993): Empirical Approaches to the Problem of Aggregation Over Individuals, *Journal of Economic Literature*, **31**(4): pp. 1827–74.

STRAATHOF, B. (2003): A note on Shannon's entropy as an index of product variety, in: *MERIT-Informatics Research Mamorandum Series, 2003-028*, Maastricht Economic Research Institute of Innovation and Technology.

TAYLOR, J. B. (1993): Discretion versus policy rules in practice, in: *Carnegie-Rochester Conference Series on Public Policy*, vol. 39, pp. 195–214.

THEIL, H. (1954): *Linear Aggregation of Economic Relations*, Amsterdam: North Holland.

TOPOROWSKI, J. and COZZI, G. (2006): The Balance Sheet Approach to Financial Crises in Emerging Markets, Economics working paper archive, Levy Economics Institute.

VAN CAMPENHOUT, J. and COVER, T. (1981): Maximum entropy and conditional probability, *IEEE Transaction on Infromation Theory*, **27**.

VAN KAMPEN, N. G. (1965): Fluctuations in non-linear systems, in: GURGES, R. R. (ed.), *Fluctuation phenomena in solids*, Academic Press New York.

WELCH, I., BRIS, A. and SCHWARTZ, A. (2003): Who Should Pay for Bankruptcy Costs?, Yale School of Management Working Papers ysm365, Yale School of Management.

WOESS, W. (1996): Catene di Markov e teoria del potenziale discreto, *Quaderni UMI n41 Pitagora Editrice Bologna.*

WRIGHT, I. (2003): The duration of recessions follows an exponential not a power law, arXiv:cond-mat/0311585v1.

ZARNOWITZ, V. (1992): *Business cycles: theory, history, indicators and forecasting*, Chicago University Press, Chicago.

Dynamische Wirtschaftstheorie / Dynamic Economic Theory

Edited by / Herausgegeben von Carl Chiarella, Peter Flaschel, Reiner Franke,
Michael Krüger, Ingrid Kubin, Thomas Lux, Willi Semmler, Peter Skott

Peter Lang · Internationaler Verlag der Wissenschaften

Michael B. Hinner (ed.)

Introduction to Business Communication

Frankfurt am Main, Berlin, Bern, Bruxelles, New York, Oxford, Wien, 2005.
420 pp., num. tab. and graf.
Freiberger Beiträge zur interkulturellen und Wirtschaftskommunikation.
A Forum for General and Intercultural Business Communication.
Edited by Michael B. Hinner. Vol. 1
ISBN 978-3-631-53710-7 · pb. € 41.90*

This series seeks to illuminate, highlight, and spotlight (intercultural) communication in the world of business. In order to conduct any business, relationships need to be established which are primarily reciprocal relationships – whether between employer and employee, or provider and customer. Since business relationships are essentially human relationships, they rely on communication. Thus, an understanding of fundamental human communication principles serves to explain, comprehend, and foster business relationships. The texts included in this book cover various topics in general and intercultural communication that have direct relevance to the world of business.

Contents: *M.B. Hinner*: Can Quality Communication Improve Business Relationships? · *A.B. VanGundy*: The Design of Ideation for New Product Ideas · *A.B. VanGundy*: Overcoming Productivity Losses in Brainstorming and Brainwriting Groups · *Ch.R. Berger*: Planning Theory and Strategic Communication · *K. Krippendorff*: Monologue, Dialogue, and Ecological Narrative · *G.J. Hofstede*: A Bridge Requires a Gap · *M.W. Lustig* / *J. Koester*: Cultural Patterns and Intercultural Communication · *L.A. Samovar* / *R.E. Porter*: Understanding Intercultural Communication · *D.D. DuFrene* / *C.M. Lehmann*: The Pursuit of Unity in Diversity · *J.N. Martin* / *R. Wuebbeler*: Contributions to International Business Practices · *E.R. McDaniel*: Culture and Communication in Japanese Organizations · *G. Hofstede*: The Universal and the Specific in 21st Century Management · *M. De Mooij*: Convergence and Divergence in Consumer Behaviour · *F. Korzenny* / *B.A. Korzenny*: A Psycho-Socio-Cultural Approach to Hispanic Market Research in the United States · and many more

Frankfurt am Main · Berlin · Bern · Bruxelles · New York · Oxford · Wien
Distribution: Verlag Peter Lang AG
Moosstr. 1, CH-2542 Pieterlen
Telefax 00 41 (0) 32 / 376 17 27

*The €-price includes German tax rate
Prices are subject to change without notice
Homepage http://www.peterlang.de